DIARY OF A

LEGAL DRUG DEALER

One Drug Rep. Dares
To Tell You the Truth

DIARY OF A

LEGAL DRUG DEALER

One Drug Rep. Dares
To Tell You the Truth

K.L. Carlson, M.B.A.

ISBN 978-0-557-11585-3

*This book is dedicated in memory of all the children
who have died because of the greed and fraud
in the pharmaceutical industry.*

Acknowledgments

Several people have been wonderfully supportive of my effort to see this book come to life. Jon and Eileen Batson kept informative articles flying my way and consoled me when writing about the harm done to children overwhelmed me to the point of nightmares. Jon, an award winning musician and author, and Eileen, writer and networking diva, have a contagious excitement about this book that I greatly appreciate.

I am grateful to Dr. Fred Baughman, neurologist and author, for his suggestions and irrepressible efforts to warn the public and government officials of the dangers of psychiatric drugs.

Special thanks to my friend David Sinnott for his editing and art work. And for his delicious cooking that kept me well fed.

Dr. Sheila Sheinberg has my everlasting gratitude for her suggestions and her encouragement.

I am grateful to Father Jim Labosky who was always willing to read new pages and provide valuable insight. His own father had been a chemical engineer in the pharmaceutical industry and had often been frustrated by the callous disregard for drug and vaccine safety.

Finally, thanks to my brother Perry for his long-distance support when I had spent too many long days focused only on writing.

Contents

PROLOGUE

In 1600 Italian philosopher Giordano Bruno was burned at the stake for promoting the astronomy theory of Polish mathematician and astronomer Nicolas Copernicus. His theory that the earth rotated on its axis and traveled around the sun defied the accepted Ptolemiac theory that had been the core of western astronomy for 1,500 years; the earth was stationary and was the center of the universe. Bruno was brought to the stake wearing an iron gag with an iron spike driven through his tongue, lest he speak any last words.

Thirty-four years after Bruno's execution, Galileo Galilie was also brought to papal trial in Rome. Under threat of torture and death he was forced to renounce all belief in Copernican theory. He fared better than Bruno. He was placed under house arrest and imprisoned within his home for the remainder of his life. Galileo, a profound genius, was then not allowed to publish or teach because his scientific studies of the solar system did not support the beliefs that helped maintain the established power structure.

Whenever the accepted beliefs are challenged, people who have power will fight the change. Figuratively speaking, people with power see themselves as the center of the universe. The western medical model revolves around the pharmaceutical industry. The medical system that has overtaken most of the world is drug-centered therefore the pharmaceutical industry is the center of the medical universe. Healthcare providers are trained to believe in this view of the medical universe and are continuously reminded to give allegiance to the drug industry.

Anyone opposing the established drug-centered medical universe is pronounced a heretic and burned at the media stake. This book is the equivalent of Copernican theory. It was written to help save lives by reminding you to not give blind and deaf allegiance to those who hold the power in the medical universe. You have the ultimate power to make your own healthcare choices.

INTRODUCTION

Life had dealt its blows, three tidal waves of grief in short succession, the deaths of both my beloved parents and a traumatic divorce. The grief was finally easing and I was determined to return to full-time work. The thought of going back to jetting around the country and world consulting left me with an empty feeling. Surely with an MBA from an excellent university I could find a position that would allow me to be at home evenings and weekends and be involved with my community. Driving past my neighbor's house, I saw Steve gardening. I knew he worked for a pharmaceutical company and on an impulse I pulled into his drive and asked if we could get together for an information interview. He graciously agreed.

That impulse led to a job as a pharmaceutical representative and an inside view of the drug industry. The total disregard for human health and human lives in the name of increased profits both astounded and enraged me.

"No spending limit," the regional sales director announced in a business meeting two months after I had completed my training. I felt like I was on an Enron train. As I looked around the room at the sixty other sale representatives, I saw dollar signs dancing in their eyes. *There is no such thing as limitless funds,* I thought as I sought out Steve, now my direct manager, during a coffee break.

"Steve, there has to be a budget. Can you give me some idea of the budget so I can plan the programs for my territory?" I asked.

Steve, a pharmaceutical veteran with more than thirty years of experience, also looked skeptical about the announcement. "That is the corporate directive, Kay. No spending limit. Just go ahead and get all the speaker programs, lunches, dinners and even breakfasts that you can get scheduled. The company is obviously determined to increase market share of product C."

"And this campaign with no spending limit is being initiated nationally with all the reps who carry product C?" I asked, my analytical mind automatically assessing the potential costs that the more than seven thousand sales representative would generate.

"I believe so. This must be an all out war to gain market share." Steve explained with concern in his eyes.

"Steve, I don't mean to sound pessimistic, only pragmatic. I've worked with many organizations around the world. I predict this corporate decision will lead to more harm than good, and it will drive up the cost of the medication. We don't need companies to take actions to drive healthcare costs up. The U.S. needs every healthcare company to remove waste and look for ways to bring the costs down while improving quality – because people's lives are the ultimate cost." I told him quietly as we stood in a corner of the room while the other reps and managers were talking excitedly as they helped themselves to the breakfast pastries and coffee. The room was electrified with people excited about spending money without limitations. I was definitely an anomaly in the group.

Steve obviously had his concerns about the announcement, but he was a company man and his loyalty to the company was evident as he tried to calm my apprehension.

"You have to let go of your education and business knowledge, Kay. This is the pharmaceutical industry. It has its own paradigm. Your job is to get doctors to write prescriptions for product C and product A." Steve tried to use the external motivator that entices most sales people, money. "The company is giving us a great opportunity with unlimited funds. Think of the bonuses you can make! In all my years in the industry, I have never been given unlimited funds to use to book programs and buy meals for offices. We must make the best of this opportunity while we have it," he said, trying to reassure me.

Two months later the company announced a price increase for product C.

Seven months later the ax fell on more than eight hundred sales representatives and managers who had been involved in product C's massive failed sales campaign. The decision apparently made so quickly by senior management that people who had been hired only weeks earlier, and were still in training, were notified that they were now out of a job. Steve and his entire team of sales people were all laid off.

Once the layoffs became public news, my telephone began to ring daily with calls from recruiters describing opportunities with other pharmaceutical companies. The free car, weeks of vacation, and financial bonuses could continue. Without a second thought I told all of the recruiters no thank you. I had seen more than enough of the fiscal irresponsibility that drives up costs and the immeasurable waste that is taken for granted in the drug industry. I had witnessed the patients sitting in doctors' waiting rooms with grocery bags full of prescription drugs, the constant push by drug companies to get physicians to put more people on more drugs, and the attitude by most physicians that drugs are the only way to treat health problems when in fact the vast majority of drugs only treat symptoms, not the underlying causes of ill health. Beyond those reasons for declining to continue to work in one of the world's most profitable industries was my strongest reason, the truths I had discovered. The pharmaceutical industry's insidious control of medical care has allowed it to become an industry of death.

I wanted to know everything about the drugs I was representing. Obviously the company's training used only the positive study results in their sales training courses. As

I researched, I uncovered facts about not only the drugs I represented, but also the deadly facts about some of the most prescribed pharmaceuticals in the world. Because the truth about the pharmaceutical industry is so frightening, and directly or indirectly affects the lives of most people, I felt an urgency to warn anyone who is open to hearing the truth about the industry that is the greatest threat to healthcare today.

My father served in the Navy during World War II. He told me the following story that had passed among the men on his ship. Bombing missions were extremely dangerous. If a plane was hit by enemy artillery the plane could burst into flames. During a flight mission over Germany, the pilot of one plane felt certain his plane's engine had been hit. Nothing happened. The flight crew completed their mission and returned to base where a maintenance crew immediately began to check and repair the plane. The flight crew was soon called back out to the hanger where their plane was being worked on. The head of maintenance showed them the several hits that the fuselage and engines had received. The flight crew was astounded. Technically they shouldn't be alive. The German munitions that had been removed were on a workbench, having been disassembled. Prisoners of war, including Russian prisoners, were forced to work in German munitions factories. All but one of the bullets had been completely empty, lacking any explosive. They were nothing but empty shells when they hit the plane. In one shell a small piece of paper with a few handwritten Russian words was found. When the Russian message was translated it read, "It is what we could do."

I wrote this book because it is what I could do.

1

SELLING LEGAL DRUGS

November 3, 2005

"Our region received a quarter of a million dollars in bonus money for product C in October!" The regional sales director's excitement was unmistakable. As I listened to his voice message, I was appalled to think what the total amount of money was that the company had spent on bonuses for one product in one month across the entire nation. The practice of giving bonuses to sales people is standard operating procedure in all U.S. pharmaceutical companies. I wondered just how many millions of dollars were doled out in the month of October for bonuses in all the pharmaceutical companies.

The ten largest pharmaceutical companies spent more than $84 billion (not a typo, it is billion) in 2005 for marketing and administration. These same ten companies spent $42 billion on research. Selling drugs is given much more emphasis than developing new lifesaving drugs and ensuring prescription drugs are safe. The emphasis on sales is the basis for the rampant fraud in the pharmaceutical industry. Companies often know their drugs are not safe, but they aren't giving out that information. They want the billions of dollars in sales.

Are the tens of millions of dollars paid annually as bonuses to drug sales people money well spent? The number of prescriptions written by doctors for a particular drug goes up and down just like the temperature. At any given time, some doctors will have written more than others. Paying bonuses to drug sales people is nothing more than lottery winnings. They are the lucky ones to have the highest numbers at a certain time. All that money paid out impacts the price of drugs. Pharmaceuticals you pay for out of your pocket, in your insurance premiums, and in your tax dollars.

I had the honor of working with and learning from the world renowned Dr. W. Edwards Deming for three years. He was the organization improvement wizard who the U.S. government asked to go to Japan when we occupied it after World War II. Dr. Deming taught Japanese companies how to ensure quality in every process to bring costs down. Japanese corporate leaders learned and applied Dr. Deming's teachings throughout their organizations and beat the economic pants off the rest of the world.

Some people mistakenly believe Dr. Deming's teachings are simply about efficiency. His profound lessons are actually a holistic approach to creating healthy organizations that always include the end users as part of the system, be they customers, patients, or students. Pharmaceutical companies are not looking out for their customers. That is why there are now **150 cases of fraud against drug companies on the Federal docket**. The number of state lawsuits against drug companies has also been increasing.

November 25, 2005

Corporate announced a price increase for product C. I was not surprised. Costs are like water. They always flow down hill creating a deep, turbulent ocean of higher prices in which consumers must try to stay afloat.

In the first quarter of 2006 pharmaceutical companies increased prices of brand-name drugs by an average of 3.9 percent. That is the highest first-quarter increase in six years. Prices of brand-name drugs increased in the U.S. an average of 6.2 percent in the twelve-month period of March 2005 through

March 2006. Over-all inflation for this same period was 3.5 percent.

Brand drug prices continued to climb in 2007. Brand name drug prices for people on Medicare Part D private insurance plans increased nearly 7 percent in the first four months of 2007. That had been the projected price increase for the entire year. The insurance premiums for the drug plans have increased 13 percent in a year, trying to keep up with the drug price increases.

Throughout this book I explain the extremely expensive, sometimes fraudulent, tactics drug companies use to sell drugs. Drug companies do not put any attention on the waste of resources due to poor management decisions because costs are passed to the payer. The payer may be an individual, an insurance company, or a government agency. You are the one at the bottom of the hill no matter which direction the costs are flowing from. You pay. The public is drowning in these costs.

Every 30 seconds someone in the U.S. declares bankruptcy because of healthcare costs. The number of U.S. families struggling to accommodate the cost of medications in their monthly budgets is an upward trend, as are taxes to pay for drugs for people on state Medicaid, federal Medicare, retired veterans, and current military.

As of June 2006 there were 100 million people qualified to receive some type of government assistance to purchase prescription drugs according to the U.S. Department of Health and Human Services. This includes the 22.5 million elderly who chose to get Medicare Part D coverage. Military retirees, no matter what their income, can receive free prescription drugs at any military base pharmacy or for only $3 per month for any prescription by mail order through the military healthcare insurance, Tricare. Your tax dollars pay the balance.

Public programs accounted for 40 percent of prescription drug spending in 2007. That is a significant jump from 28 percent in 2005. It is not an accident. The drug industry has been lobbying the federal and state governments to establish more ways for public money to flow into drug industry bank accounts.

"Always ask for the Medicaid business." Managers continuously reminded me and the other drug representatives. It was a corporate directive. People in our state aid program paid only three dollars for any prescription, which meant they would probably continue to take the drug for years or even the rest of their life. That is why all drug companies are targeting Medicare and Medicaid patients. The drug industry has a well-oiled machine in place to grab increasing amounts of money from Medicare and Medicaid. Some state Medicaid programs have gone bankrupt because of drug expenses while many patients have suffered additional health problems because of drug side effects.

The pharmaceutical industry has a death-grip on the U.S. medical and government systems. This book presents the evidence so you can decide for yourself.

The U.S. spent an amazing $252 billion for prescription drugs in 2006. The U.S. is the most drug-consuming nation on earth. And the number of prescription drugs per capita is a rising trend. In 2005 the U.S. per capita of prescriptions was 12.3. This means when the total number of prescriptions is divided by the total U.S. population the number of prescriptions is more than 12. That is an increase from 7.9 prescriptions per capita in 1994. Americans are gulping down huge amounts of drugs daily. And yet a 2006 Harvard University study found that life style is the most common reason for life expectancy in the U.S. But instead of educating patients to improve life choices, doctors are pushed by the pharmaceutical industry to write prescriptions for symptoms. The naïve public allows the drug-centered medical system to put them on multi-drug therapy, often causing side effects leading to more drugs.

At the beginning of 2007 46.6 million Americans were without health insurance. That is an additional 7.6 million people in just five years. At the beginning of 2002 nearly 39 million Americans were living with the stress of possible financial ruin because they could not afford health insurance.

The rising cost of drugs feeds into the rising cost of health insurance. Health insurance premiums are affected by both the high cost of pharmaceuticals and the number of people taking them. And as FDA records confirm, drug companies are sometimes dishonest with doctors and the public in order to get more people to buy brand name drugs.

December 4, 2005

"Convince doctors that product A is excellent for treating pre-hypertension." The email from corporate marketing instructed us. I read it and shook my head with disgust. Apparently the company wanted people without health problems to be encouraged to take synthetic drugs.

Marketing drugs to the healthy is one of the great hoaxes that the drug industry has perfected. It is a major reason why so many relatively young Americans are taking prescription medications. According to AARP, "most people over the age of 45 in the U.S. take an average of four prescription medications daily."

In their well-researched book, *Selling Sickness*, authors Ray Moynihan and Alan Cassels raise the important issue of pharmaceutical companies creating the concept of diseases in order to increase sales. Drug companies, often through front groups, have advertised to convince the public that as many as one in five people are depressed not just going through the natural ups and downs of life; that up to 25 percent of children have the Attention Deficit Disorder, not that they are naturally active children hyped up on too much sugar, caffeine, and processed food, or are suffering from lead poisoning; and menopause is a health problem of estrogen deficiency, not a naturally occurring process that is actually an imbalance of estrogen to insufficient progesterone. The list goes on and on. Drug companies spend millions of dollars contriving and naming health conditions with the help of medical associations that they financially support. Then they spend billions of dollars marketing the newly created health concerns and their pills that are the perfect answer. All the spending is for the purpose of getting more people to buy more prescription drugs. It is not about health. It is about money.

Perhaps the older generation is not so susceptible to the omnipresent pharmaceutical advertising because approximately 10 percent of Americans 65 years or older do not take any prescription medications on a regular basis.

I have two octogenarian friends who do not take any medications. Irene celebrated her 90th birthday in February 2009. She takes no medications. She lived in her own home in Minnesota until the age of 89, remaining physically and socially active. Irene's motto is "You can't beat fun." She happily admits she has not been to a doctor's office in three years.

Another friend, Marvin, is 82 years old and takes no medications. He stays active on his horse farm and doesn't flinch at every little pain. Marvin doesn't believe in depending on pills.

Merck would "sell to everyone" if former CEO Henry Gadsden had his way. He told Fortune magazine thirty years ago his dream was to make drugs for healthy people. He wanted drugs to sell like chewing gum.

The philosophy of strong leaders like Gadsden can influence a company and an entire industry long after the man is gone. His wish marked the beginning of the drastic change in the aim of the pharmaceutical industry. The great shift from an industry that was focused on creating lifesaving medicines for the truly ill, to an industry that is intent on convincing constantly growing numbers of people that prescription drugs will enhance their lives. Drug companies would be happy to have every man, woman and child taking brand name drugs. What got lost in this race to always sell more drugs is that prescription drugs should be medications only used for special reasons. Drugs should not be treated as consumer products like candy.

2

THE HISTORY OF MARKETING PHARMACEUTICALS

As of June 2007 there were 100,000 pharmaceutical representatives in the U.S. calling on physicians. Physicians' offices are overwhelmed with the number of patients to be seen daily and the administrative work required in today's healthcare environment. Yet the ranks of pharmaceutical representatives are growing monthly. I was contacted in May and June 2006 by eight recruiters representing pharmaceutical companies. Seven were U.S. drug companies and one was a foreign drug firm that had just received FDA approval for its drug. Some were large companies and some were small start-up pharmaceutical firms. All drug companies are locked into "this is the way we have always done it" mode. Door-to-door selling to doctors is how business has been conducted for half a century.

The pharmaceutical industry has a vast resource-devouring behemoth on their hands. How did the method of selling door-to-door to physicians develop in the United States? I had the opportunity to interview two men, John and Ken, who have each been pharmaceutical representatives for almost five decades. They described the changes in selling prescription drugs to doctors; changes they both posed concerns about. I found these men to be dedicated to the principle that drugs should be medicines for the truly ill. And that drugs should heal, not do harm.

"Very technical and detailed," were how John and Ken described the discussions representatives had with physicians in the 1960's and early 1970's. It was a time when pharmaceutical representatives were trusted and respected by physicians and their staff. The pharmaceutical companies doing business in the 1960's hired a small sales force to market their products. Every company had only one or two representatives for an entire state. A representative would know all the company's products and call on physicians of all specialties. Doctors saw only one or two pharmaceutical representatives in an entire week. One elderly physician practicing in a rural small town told me he saw only two drug representatives per month during the 1960's. Therefore the doctors took time to discuss products with representatives. The discussion was about use of the drugs as true medications to be used at appropriate times. "Doctors expected us to know our products inside and out. Doctors trusted and respected us then. It is completely different today." John sadly told me.

Pharmaceutical companies hired representatives who were either degreed pharmacists or were certified by the Certified Medical Representative Institute in Roanoke, Virginia. The institute offers a training program designed for pharmaceutical representatives to gain knowledge in anatomy, pharmacology, and adverse events (drug side effects). All pharmaceutical firms required their representatives to complete the institute's program if they were not pharmacists. The program took two to three years to complete on a part-time basis for working drug representatives.

Compare that with today's hiring practices:

May 16, 2005

The training room was filled with forty-five people, most of them young and attractive. Outstanding among the women in the group was a living Barbie Doll. We were instructed to introduce ourselves to the group and give a brief history. "My name is Cheryl (not her real name). Among my past jobs I was hired by Mattel Toys to perform as Barbie." I wasn't surprised that "Barbie" was hired as a drug rep. What male physician was going to refuse to spend time with her?

There is at least one drug company that actively recruits college cheerleaders for their sales force. Current hiring practices differ dramatically from the past. Any four-year degree is acceptable; no science education is needed. I met representatives with diverse backgrounds, from former teachers to aerobics instructors.

Pharmaceutical companies provide only two to four weeks of training for each drug a representative will be selling. The emphasis of the drug companies' training is about selling; how to get the doctors' attention and try to get as much time as possible with every doctor. Today it is about selling, selling, selling. Just get the physicians to write more prescriptions for the products you represent. If the doctors are concerned about side effects, distract them. Get the doctors to push drugs.

John described to me the early days of calling on physicians. There was a method called "smoke stacking". When a representative arrived in a town that he (99% were men) did not know well, he would go to a public phone booth and tear out the yellow pages for physicians. He would then begin to locate physician offices from the addresses from the phone book pages. Representatives would also call on pharmacists and learn which physicians were prescribing the products the representative carried and how much of the drug had been needed to fill prescriptions in the past month.

There were no computer databases in those days tracking every prescription that had been written by every doctor. Today every drug company uses these databases. Drug samples could legally be left with pharmacists. Legal restrictions no longer allow sample distribution to pharmacists. The current federal regulations allow drug samples to be left with healthcare providers who can write prescriptions as defined by state laws. In some states nurse practitioners and physician assistants can write prescriptions, so they are called on by pharmaceutical representatives and receive drug samples.

In the 1960's and early 1970's representatives spoke with physicians using the package inserts (P.I.). There was no expensive, glossy Madison Avenue marketing material. Package inserts are available for every drug. The P.I. provides details of the drug including warnings, cautions, possible side effects, and the recommended starting dose. The P.I. is small print without attention grabbing color photographs or multicolored graphs. The P.I. gives the positive and negative (though not always a complete disclosure) information about the drug.

"There used to be tight limits on expenses." Ken explained. Drug sales people were paid a modest salary and no bonuses. Most drug companies reimbursed the representatives for miles driven rather than providing company cars. There was no entertainment budget to take physicians out to dinner or to cater lunches into the entire office. There was a minimal "burger budget" that allowed representatives to occasionally take a doctor to a diner for a burger lunch to provide more discussion time.

John told me sadly, "In those days we made a living but not a lot of money. People respected us. Now when you tell people you are a drug rep. their first reaction is 'so you make big bucks.' Anything to do with pharmaceuticals today is just seen by the public as big dollar signs. I don't like that image, but I can understand why the public has it."

In the 1980's there were three marketing changes that significantly increased the percent of sales price used to market each drug. The three changes were: 1) expensive entertainment budgets that had not existed in the previous decades; 2) increased number of employees selling or supporting sales of the products; and 3) paying bonuses to sales representatives and managers.

The percent of each product's sale price that was needed to cover marketing costs increased exponentially.

"Some representatives did nothing but play golf throughout the 1980's." John described the "easy money" of the era. Paying to play golf with physicians was an acceptable selling technique. The 1980's were when significant changes in marketing drugs to doctors began. Merck and Pfizer, the largest drug companies, began to pay quarterly bonuses to their sales representatives. The inconspicuous "burger budget" was replaced with significant entertainment budgets. By the late 1980's all pharmaceutical firms allocated sizeable entertainment budgets to attempt to influence physicians. This marks the transition point from the primary costs of pharmaceutical products being research and manufacturing to marketing and administration being the greatest expenses of drugs.

In the mid-to-late 1980's pharmaceutical firms began courting physicians with extravagant entertainment. Cruises, island vacations, and golf outings to the best resorts are just a few examples of the expensive entertainment used by all the major pharmaceutical companies to persuade physicians to prescribe their products. Physicians and their spouses were lavished with costly entertainment and trips; with the constant theme – write more prescriptions for our drugs.

Allocating large marketing budgets to entertain prospective customers is a type of sales strategy that is used by all types of companies. But who is really the "customer" of pharmaceutical products? The healthcare system does not fit the traditional business model. Ethics are not questioned when expensive marketing causes price increases of products consumers can make a choice to buy or not buy, although it is still a wasteful practice. However, when the products are medications that people's lives may depend upon, wasteful, expensive marketing practices that drive up the price do seem unethical. The cost of medications is the number one reason for patient noncompliance, meaning the patient does not take a medication that his/her physician has advised taking. The climbing cost of Medicare and Medicaid paid by our tax dollars should also have any ethical drug company looking for ways to decrease the cost of their pharmaceutical products. Instead, drug companies look for ways to leech public healthcare money.

And physicians are being encouraged to prescribe drugs when non-drug therapy would be more beneficial. The U.S. population has become the most drugged and least healthy western population. Americans consume almost half of the total world sales of prescription drugs annually.

John explained the strong influence the largest drug companies have on the industry. In the 1980's Merck initiated a new marketing strategy for their products. The "shared voice" strategy is based on the theory prescription writing for a drug will increase if healthcare providers are called on frequently (sometimes daily). This strategy required a substantial increase in the number of sales representatives. It seemed to work for Merck and all pharmaceutical companies followed the leader, expanding their sales forces to make more frequent sales calls on all doctors. Along with the increased number of sales representatives the number of managers and administrators at each company were also increased to support the expanded sales forces.

"Since Merck was the largest, all the other firms thought it was best to copy anything Merck did," John explained. Another contributing factor to the increase in sales force size was the increased number of pharmaceutical products. The "shared voice" strategy transitioned the sales force from familiarity with all of a company's products to a sales force of limited focus. Merck's sales force morphed from generalists to specialists. Representatives were hired to carry only one or two products instead of the company's entire line of drugs. All pharmaceutical firms followed Merck's lead.

"Some physicians actually demanded expensive trips." Ken shook his head remembering. The pharmaceutical companies had shown themselves to have deep pockets so some physicians wanted to take full advantage. Neither group considered the money had to come from the patients who needed to take the medications, or from tax dollars everyone paid. The extravagant courting of physicians by drug companies reached its apex in the early 1990's.

"In the mid-1990's pushback from the government began with a bill proposed by Congressman Bingell of Michigan," John explained. Drug companies did not want to be restricted by government controls. The industry countered the government's restrictive threats by proposing to the government the pharmaceutical industry would create regulating guidelines that pharmaceutical companies would voluntarily follow. The U.S. federal government yielded to the pressure of the wealthy, influential pharmaceutical industry – so now the fox guards the hen house. The major U.S. pharmaceutical firms created the Pharma Code to regulate their industry. The Code establishes limits on the amount that can be spent on lunches or dinners. It prohibits inclusion of spouses, unless the spouse is also a healthcare provider. It restricts entertainment: no more trips, golfing, or cruises. The Pharma Code remains voluntary. Some companies strictly follow it. Other companies still believe expensive entertainment of physicians is key to expanding market share.

Direct To Consumer Advertising

July 8, 2005

I rolled into the smallest town I visited; so small there is not even a traffic light in the town. I saw on my computer that another sales person in my group had been here to call on the one physician in town only two days ago. I thought it was unnecessary for another visit so soon, but corporate assigned the number of visits I must make to each physician. I took a seat in the unsophisticated, somewhat worn waiting room where a young mother held a small child in her lap. Dr. B. looked to be about sixty. He had me step into his personal office. Just out of training, I introduced myself. Dr. B. looked at me solemnly.

"These aren't cereals or ice cream." Dr. B. told me in frustration. "I have patients who have never completed high school self-diagnosing according to drug ads they see on television." He spent another minute of his precious time describing his growing impatience with patients who come to his office demanding a prescription for a drug they had seen advertised, including a drug I represented.

In the late 1990's the U.S. federal government again showed disregard for the public's best interest and softened the regulations for pharmaceutical advertising direct to the public. Studies have found if patients specify a brand of drug to their physician, more than half the time the physician will write the prescription for the requested brand. Dr. B. was only one of several doctors who admitted to me their frustration with patients demanding prescriptions for drugs they had seen advertised. Former Merck CEO Henry Gadsden's dream of selling prescription drugs to everyone is coming true in the U.S.

The U.S. and New Zealand are the only countries that allow drugs to be marketed directly to consumers. The drug companies have been pushing to advertise in other countries but so far the governments of all other countries have prudently resisted. People in these countries do get prescriptions from their physicians for appropriate medications, but are much less likely to be over-medicated or unnecessarily medicated.

Drug advertising often plays to people's fears so that even healthy people believe they would be better off taking a drug. As advertising agencies are keenly aware, people's choices are directed by fears or desires, so advertising specialists create drug ads to play on people's fears or their desire to be happy.

A good example is the current emphasis on low cholesterol. Drug advertising would have us believe that by taking a statin drug to lower cholesterol, you have less chance of suffering a heart attack (fear of death). That is not what the research shows. Clinical trials have NEVER shown statins reduce heart attacks or death in women. That is only one example of how the direct to consumer advertising of drugs can mislead the public.

Of course the intent of the expensive advertising has nothing to do with improving health. The intent of the advertising is to sell drugs. And it is working. Drug companies spent $4.2 billion for television and print ads in 2005. The **U.S. population purchased 3.6 billion prescriptions in 2005**, an increase of 71 percent from 1994 to 2005.

When you see pharmaceutical ads just remember, it is not about science, it is about moving money from your pocket into drug companies' pockets.

3

EXPENSIVE DOOR-TO-DOOR SELLING TO DOCTORS

July 12, 2005

It was an overcast morning in the small North Carolina town with a population of about 6000. There were only three physicians I had to see in this town. Still new to the large territory I had to travel, I was finding the offices of professionals I was assigned to call on. I parked the company car across the street from Dr. N's office and got out my detailing bag, full of colorful selling aids and gifts for the office. Looking at the full parking lot I thought he must have many patients this morning. Instead of many patients, what I found in his waiting room were ten other pharmaceutical representatives.

Every week at least one, sometimes two or three representatives from the same drug company with the same medications will attempt to see the same physicians. This pattern is repeated every week of every month. That is the marketing strategy of pharmaceutical companies; get drug sales people in doctors' offices often saying the name of the product. It is a very expensive strategy. Costs are like water. They always flow down hill. You are at the bottom of this ever-growing flood of costs. An ocean of costs is rising to drown you.

"Fifteen to twenty-five representatives a day is normal." Explained my manager, Steve. Primary care physicians, those in internal medicine and family care, can expect to have 15 to 25 drug sales people attempt to see them every day. As an organizational improvement professional I was horrified.

I considered if each representative took just two minutes of a doctor's time it still added up to thirty to fifty minutes every day. That is valuable time to be hearing a sales pitch that a physician has been hearing every week for sometimes several years.

Imagine having your work interrupted fifteen to twenty-five times every day to hear a sales pitch for your printer cartridge, or printer paper, or light bulbs. My point is healthcare providers know the medications just as you know products you work with every day.

Why don't the physicians just say no? Some do. A small but growing percent of physicians now refuse to see drug representatives. Many physicians see drug reps only to sign for the free drug samples because the law requires it. No drug samples can be left at an office without the doctor's signature. Some offices only allow drug sales people access on certain days or certain hours, trying to minimize the wasted time yet afraid to say no completely because they want the free drug samples.

July 17, 2005

I stepped from the humid heat into the air-conditioned office of Dr. J. This was my first visit to his office. I approached the reception desk and introduced myself.

"Dr. J. will only see drug representatives if they bring lunch in for the office," she explained. I was appalled but asked when the earliest opening was available, disappointed to discover it was two months in the future. The company expected me to see this physician at least twice a month. The four other sales people in my district would also call on him with the same drugs.

In large multi-physician practices it is common to provide lunch for 50 to 70 people. Often the physicians will make up a plate and return to their offices so they can continue work on the mountains of paper work mandatory in today's medical environment. It was common for me to spend hundreds of dollars for a lunch and still have less than two minutes with the physicians. By anyone's standards, not cost effective. But the company expected us to use lunches as a way to gain doctors' attention and appreciation. The management mentality from the 1980's and 1990's remains intact today in all drug companies.

There is a backlash occurring as physician groups such as the New York-based No Free Lunch and the American Medical Student Association campaign to encourage physicians to not be subtly bribed by pharmaceutical companies. Healthcare should be about healthcare, not pumping pills at every patient entering the office. Keep in mind the basic aim of the pharmaceutical industry is to get as many people taking as many brand prescription drugs as they possibly can. That includes you, your child, your teenager, your parents and everyone you know. The pharmaceutical industry's mantra is "More people on more pills."

As a drug representative I was assigned to call on physicians who had been called on for years by representatives carrying the same products from the company. All the major pharmaceutical firms maintain large sales forces in order to make frequent calls on physicians. And as representatives are hired, more managers and administrators are also hired.

The U.S. is now inundated with a ratio of one drug sales person for every nine doctors. More drug sales representatives are hired every time the FDA approves a new drug. When the trend is projected out over the next five years the number of pharmaceutical representatives calling on physicians daily becomes absolutely stifling. Between 1996 and 2006 there was a tripling of the number of drug representatives to 100,000 according to Verispan, a healthcare data company. Imagine if there is a tripling of the number between 2006 and 2016! If the trend continues that could happen. Even as I write this, the third and fifth largest Japanese pharmaceutical firms have partnered to market recently FDA approved drugs in the U.S. and are hiring a sales force to make door-to-door sales calls to physicians.

Plethora of Numbers

As mentioned earlier, in the 1960's and 1970's a drug representative would talk to the local pharmacist to learn what drugs the local physicians were frequently prescribing. This locally gathered data helped the sales representative to determine what products he should discuss with each physician. The data, the number of prescriptions, was not used by corporate as a carrot or a club against the sales force. Now data are used very differently. Numbers are clubs and carrots.

In today's computer-driven world every prescription written can technically be collected from pharmacies. Not all pharmacies participate but most of the major chains do sell prescription data to data companies. These companies then sell the data to pharmaceutical firms. The data indicates how many prescriptions of any brand of drug every physician has written weekly and monthly. There is no patient information involved. It is strictly numbers of prescriptions.

How do drug companies use these numbers? The companies want to know what class of drugs physicians are prescribing and what percent of the market share their products have versus every competitor. The market share numbers are tracked weekly. What are drug classes? It is the general type of drug. For example statin is a class of drug for lowering LDL cholesterol levels. There are several brand names of statins and there will soon be a generic. Any physician who writes many prescriptions for statins in a month is a target for all the pharmaceutical companies with a statin product. The market share numbers are available to the sales forces on their company computers. Drug firms expect their sales representatives to apply pressure to any physician who is prescribing the class of drugs the sales people carry. "Doctor, can I count on you to prescribe product C for any patients with X health issue?"

"Always ask for the business." This message is emphasized throughout the training and at every sales meeting. Sales messages are droned into sales peoples' heads: "So doctor you will write more prescriptions for product C won't you?" Or "Doctor, you will prescribe product A for your heart failure patients, won't you." Or "Doctor, I can count on you to prescribe product C for ten new patients this week, right?" All the emphasis was on continually increasing the prescription numbers. Drug companies judge the success of their sales people from the numbers collected from the data companies. Drug firms expect the numbers to be on a continuous increase.

As Dr. W. Edwards Deming emphasized, it is easy to misuse numbers. Data should be looked at over time and used to improve the processes, not to manage the people. Some of the things ignored by drug companies that are factors in prescription data: 1) Not all pharmacies sell data. Walmart pharmacies do not and in some rural areas Walmart fills the majority of prescriptions. The sales force is still held accountable for reported low prescription numbers. 2) If a physician is on vacation, no prescriptions are written. Why punish the sales person because the numbers decreased? 3) If a new physician moves into an area, prescription numbers may change. Why give a bonus to the sales representative for the area? These are just a few of the examples of situations not controlled by drug representatives that are ignored by drug corporations as they manage people by numbers. Managing people by numbers creates waste of resources and increases costs.

Additional Costs of Door-to-Door Selling

One of the major expenses involved in door-to-door selling of pharmaceuticals is transportation. All the major drug companies provide cars and gas credit cards for their sales people and sales managers. Some of the smaller drug firms pay their salespeople a monthly stipend for use of personal cars and provide gas credit cards. The miles driven per week will vary greatly depending on the drug representative's territory. I drove an average of 750 miles per week. Hundreds of gallons of gas consumed in order to give out free drug samples and free office gifts every month. Not to mention the unhealthy impact to the environment. Day after day I would arrive home and think "what a terrible waste." I find this door-to-door selling process depressingly out of date, just like door-to-door selling to houses became outdated. I considered the hundreds of miles I drove every week (not in a hybrid car) representing two medications that have been on the market for three and four years. The four other sales people on my team were also driving all those miles and seeing the same healthcare providers about the same drugs and usually the time with a physician was ten seconds for a signature to receive samples because physicians are busy and they have had the slick Madison Avenue marketing material shoved in their faces too many times.

Every car is packed with expensive product brochures and free gifts for the medical offices. The free gifts have gone well beyond pens. Some offices become absolutely cluttered with gadgets toting the brand names of drugs. The Pharma Code specifies that any promotional gift must be worth less than $100 and must be medical office oriented. The drug company I worked for provided an entire database of more than one hundred items from which the sales force could order. It was common for me to give several items at every office I visited. In some of the small rural towns where a high percentage of the patients are Medicaid participants, doctors' offices often depend on the gifts from pharmaceutical sales people because these offices have limited income and operate on a tight budget. But is it good to have physicians dependent on pharmaceutical companies for office items? Those "free gifts" end up in the cost of the pills.

August 4, 2005

Another extremely hot day in a succession of days with the heat index over 110 degrees. The National Weather Service warned that the heat inside a car will quickly climb to 150 degrees. What did that heat do to drug samples, I wondered. I felt fortunate I did not have to physically carry samples with me. The physicians sign on a small computer I carried and the samples are shipped to their offices. As I sat in a waiting room glad to be in a cool office, I watched another representative struggle through the door with a large bag of samples. She took a chair across from me to wait her turn. I asked her if she had to remove the samples from her car at the end of every day. Yes, she responded, she was supposed to according to her company's policies. However it was a time consuming and laborious process that was difficult to face after a day of working in the heat, so she only removed the samples from her car once or twice a week. Her reply made me wonder if the large quantities of samples shipped to sales people is always stored in air- conditioned storage or left in a salesperson's hot garage. No doubt heat has a deleterious effect on some drugs.

August 23, 2005

As I checked the expiration dates on the samples at Dr. N.'s office I found seventeen packages of product A that were out of date. Each sample was one week of medication. The cash price (no insurance) for a month's supply of product A was approximately $85. And here in one office was 17 weeks of a blood pressure medication that would be thrown out. This was not an unusual occurrence. I found expired samples in offices every month.

Pharmaceutical companies spent almost $16 billion in free drug samples in 2005. All medical samples have expiration dates. The office staff sometimes goes through and disposes of outdated samples, but in today's busy office environments it is often left to the drug representatives to check for expired samples. The law does not allow pharmaceutical representatives to throw out any samples. They can set the samples apart from the good samples and notify the office staff so the expired samples are properly disposed. Every month throughout the U.S. thousands of samples of medications are thrown out because they have expired. It cost to produce the medications, it cost to ship them, and it cost to pay the representative to take them into the office. Yet every month thousands of samples of medication are thrown out because they were left on the shelf rather than given to patients. That is expensive garbage.

Why do drug representatives leave too many samples? There are several reasons. Pharmaceutical companies want their samples in sample closets because if doctors give out a sample, usually a prescription goes with it. Some doctors will tell representatives, "I will write prescriptions for your product if you keep my office well sampled." I have thrown out dozens of samples in offices where the doctor has been telling representatives that line for years.

Another reason for over sampling is that pharmaceutical companies assign their representatives a list of physicians and how many times in a quarter or trimester they are supposed to see each doctor. This is another example of managing people by numbers. It is very top down communication and therefore is extremely inefficient. In fact, I was constantly assigned to visit doctors who were dead, or had been retired for years, or had left the area. I would have to go through a formal process to remove the names every trimester. Then the next trimester they would be back on my list again. Waste of resources, including employees' time, drives up costs, and you know who ends up paying for those costs.

A representative has only one way to prove the physician was seen, the doctor's signature for samples. In 1982 federal law was passed requiring the prescribing healthcare provider must sign for samples, whether the samples are left by the representatives or the samples are shipped directly from the drug company to the doctor's office. So it behooves the representative to get a signature proving the healthcare provider was seen, if only for five seconds to get a signature for samples.

There may be much less expensive and safer methods to provide samples that do not require door-to-door selling. One possibility is certificates that patients could be given to take to a pharmacy and get a small supply for free in order to try the medication before filling a prescription. This method would eliminate the expensive, wasteful method of maintaining drug samples in physicians' offices. Doctors could request and print sample certificates online from secure websites. A method like this would also eliminate the waste of expired samples since physicians could request only the medications they feel comfortable prescribing. It is out-of-the-box thinking that could save billions of dollars every year because it would eliminate door-to-door selling. The savings could be passed on in terms of lower drug prices for the payers of medications, providing some relief from the escalating cost of healthcare.

Pharmaceuticals Are In Drinking Water

Outdated drugs are being disposed of in dangerous ways. *USA Today* reported in September 2008 that hospitals and long term care facilities were dumping an estimated **250 million pounds of pharmaceutical drugs into public sewer systems every year.** These dumped drugs are ending up in drinking water because the current facilities are not capable of removing the drugs from the water. The article states, "Commonplace presence of minute concentrations of pharmaceuticals in the nation's drinking water supplies are affecting at least 46 million Americans." European studies have also found pharmaceutical drugs in drinking water.

A national survey of physician-industry relationships found 94 percent of physicians have some type of relationship with the drug industry. Doctors could help bring about innovative solutions to expensive door-to-door selling by not permitting drug salespeople to call on them. If a majority of doctors and hospitals would stop seeing drug sales people, the drug companies would be forced to reconsider using door-to-door selling.

Newsweek reported in October 2007 that the tide may be turning as the number of physicians saying no to drug representatives is slowly increasing. The American Medical Student Association collected twice as many pledges from medical students in 2007 as in 2006. The pledge is that after graduation from medical school the new physicians will refuse drug company gifts and drug representative visits.

More Expensive Garbage

September 12, 2005

The company's marketing department had devised a promotional item for product C. As soon as I saw how much time it required for physicians or their staff to explain to patients, I knew these items were going to gather dust in sample closets.

Drug companies often design promotional items to be given out to patients. Sometimes it is a simple coupon. At other times it is an involved gimmick that neither patients nor physicians want to fool with. Usually these ideas are not prototyped and tested in a small region to ensure that the promotional item is actually used. Millions of dollars are spent on promotional materials that gather dust in offices until the office staff needs space and tosses out the promotional materials or the materials expire and are thrown out. Like expired samples, this is expensive garbage.

What seems like a good idea at corporate headquarters may be cumbersome or time consuming in busy medical practices. Medical offices are inundated with record keeping and administrative work. Anything that is outside of normal office routines is going to be ignored by physicians and office staff. Even coupons that save patients money when prescriptions are filled are often left on the shelf. The pace of most medical offices does not allow extra moments to get the correct coupon and tell patients how to use it.

September 23, 2005

"Unlimited spending! Schedule all the programs you can." That was the management directive announced at the regional business meeting. The Enron train was roaring down the tracks and the company expected everyone to be on board.

Weekly emails from management began to arrive showing every representative's number of scheduled programs, creating a competitive atmosphere; as though the people with the least number of programs scheduled should feel guilty. Daily voice mail messages from the district manager announcing to the team which representatives had scheduled new speaker programs gave the situation a television game show atmosphere. Representatives competed to get the most notable and expensive speakers scheduled. The best restaurants were booked for dinner programs in the hope of luring physicians to attend. It was a costly strategy for medications that had been on the market for three and four years. The company's theory and hope was all these programs would lead to growth of market share for two products, especially product C. The company did not run a prototype by testing this theory in one or two areas of the country to check the results. The plan was opened all at once across the entire nation.

At times I provided breakfast at one office, lunch at another office, and then a speaker dinner program at a restaurant all in one day. I spent $1000 to $6000 a week for all the meals and speaker programs. I was only one of 7000 in the company's sales force.

4

TRICKS TO PAY DOCTORS
TO WRITE PRESCRIPTIONS

October 4, 2005

Dr. O. greeted me and reached for the small hand-held computer to sign for the samples that would be shipped to his office. "I want to be a speaker for product A. Will you please find out when the training will be so I can put it on my schedule?"

Speaker programs are a marketing method commonly used since the year 2000. Physicians chosen to be certified speakers for a brand name drug receive brief training, usually at a very desirable location, for the slide presentation created and approved by the drug company's marketing department and based on the Pharma Code.

Certified speakers then make presentations at lunch and dinner programs scheduled by drug representatives. These paid professionals help convince other physicians a particular brand of drug is good to prescribe. Or they may promote new guidelines, such as for cholesterol, that increases the market for drugs by millions of people. Physician speakers are compensated in a range from $500 to $2500 per program.

"I'll let my manager know your interest." I answered politely. But my mind was reviewing the several unattended dinners featuring Dr. O. as the speaker for product C. Just as in any profession physicians' skills in their work is not a reflection of speaking ability. Dr. O. was an ineffectual speaker and other physicians were not attracted to come to dinner programs featuring him.

Imagine you are a drug sales representative and a physician in your territory who writes many prescriptions for your products tells you he wants to be certified to be a speaker for a product. This physician has been a speaker for another of your products but consistently no one has attended his programs. You want to keep him happy and the company is paying you to get physicians to write prescriptions. Are you going to tell him no?

This was the dilemma for my team. In the end our managers agreed with the two team members who were adamant Dr. O. be certified to speak for yet another of our products to "ensure he keeps prescribing our products."

There is a federal anti-kickback law prohibiting payment by pharmaceutical firms to physicians for writing prescriptions. The experience with Dr. O. proved to me that there are ways to get around the law. While drug companies require their salespeople to take training about federal laws and regulations, corporate policies governing employee compensation and even the risk of job termination are at times in direct opposition to these very laws and regulations.

Former Merck regional sales manager, Gene Carbona, told the *New York Times* that the only thing the company considered when selecting physicians to provide presentations was "the volume or potential volume of prescribing that the doctor could do." This is true of all pharmaceutical companies.

October 10, 2005

"Dr. L. is now trained to speak for product A and wants to do a dinner program," a team member informed me.

"But we just had a dinner program with a specialist as the speaker. Dr. L. is not a specialist and who is going to come out to hear a local primary care physician about the same product?" I expressed relevant concerns. My teammate's response to the situation was again a method to appease the physician who wanted to be paid to speak. "Since Dr. L. has a large practice with several healthcare providers, I'm going to suggest we do a dinner program for his office. I'll find out what restaurant they want to use and pick up their dinner orders. We can set it up in their office conference room."

There is nothing legally wrong with this suggestion. For the team member it counted towards his monthly total programs the company was pushing us to schedule. It also kept Dr. L. happy because he would receive the speaker compensation. And if it kept him happy with the company then the hope was he would increase his prescription writing for the company's products.

The fear of drug representatives is that a physician will decrease the prescriptions written for their products. Dr. L. is also a paid speaker for a competitor drug company for a drug in the same class as our product. This is analogous to a star tennis player promoting two brands of tennis rackets. It put pressure on us to ensure he was happy or he might write more prescriptions for the competitor's drug because they paid him to represent their product. Essentially, both companies pay him to write prescriptions.

If a local practicing physician has the potential to make thousands of dollars a year as a pharmaceutical speaker will it influence his or her tendency to prescribe drugs to more patients? Doctors know drug companies routinely watch prescription numbers. Research shows that physicians who have any type of relationship with drug companies write more prescriptions than doctors who do not. It raises the possibility some patients are being put on drugs that offer them questionable benefit while exposing them to the risk of serious side effects.

Some Physicians Can Legally Make Millions from Drugs

The anti-kickback law does not apply to drugs that must be administered to patients in a doctor's office. The *New York Times* reported on May 9, 2007 that one practice of six oncologists (cancer specialists) was paid $2.7 million in 2006 by two drug companies. Hundreds of millions of dollars have been paid out to physicians across the country. Paying rebates to physicians is a legal manipulation to get around fraudulent and false claims to federal and state healthcare programs.

The payments were made for giving patients a class of anemia drug known as EPO. Just as chemotherapies are drugs given to patients in clinical settings, the anemia drug is administered in physicians' offices, cancer clinics, and dialysis centers. Physicians purchase the drugs from the drug companies then receive a rebate. Physicians may charge the full markup price to Medicare or private insurers for the drugs.

The more EPO drugs a doctor purchases the larger the rebate given by the companies. The rebate program appears to impact patient care. The average dose of the anemia drugs given to dialysis patients in the U.S. has tripled since 1991. Dialysis patients in Europe where there are no rebates receive half the dose amount American patients receive. Cancer patients in the U.S. are three times as likely to be given the anemia drugs as cancer patients in Europe.

There is no evidence the anemia medications either improved quality of life in cancer patients or extended their survival, according to a report by FDA scientists released May 8, 2007. Federal statistics also show dialysis patients do not have an improved survival rate with increased use of the anemia drugs.

EPO drugs are useful in very specific types of patients, those with severe anemia such as some patients with chronic kidney disease. There is no evidence EPO's make a difference in patients with moderate anemia. The annual death rate of dialysis patients has remained at about 23 percent, unchanged since EPO's were introduced in the early 1990's.

There are significant health risks with EPO drugs. More than ten years ago there was clinical evidence EPO's increase the risk of heart attacks when used in doses that increased hemoglobin levels above ten. Clinical trials with cancer patients in 2003 showed the drugs can worsen cancer or hasten death.

In March of 2007 the FDA required EPO drug labels to carry stronger warnings of the cancer and cardiovascular risks with higher doses.

The drug companies marketing actions are paying U.S. doctors to put patients on higher, more dangerous doses. Better bottom line for drug companies. Better bottom line for physicians. End-of-the-line for many patients.

5

GUIDELINES ARE MADE FOR DRUG SALES NOT HEALTH

January 23, 2006

"My father died of a heart attack at age 68 and his cholesterol levels were fine." Susan, a nurse practitioner in a large clinic, explained to me why she was not impressed with cholesterol guidelines.

My company's training had emphasized pushing the new lower cholesterol guidelines to impress upon healthcare providers that more patients should be prescribed the company's statin. I had spent enough time in the organization that I had begun to doubt they had given their sales force all the studies about statin drugs. I decided to do my own research and what I discovered infuriated me. If I hadn't received notice that we were all to be laid off, I would have quit just to be out of the deception.

"The diet-heart idea (the notion that saturated fats and cholesterol cause heart disease) is the greatest scientific deception of our times. This idea has been repeatedly shown to be wrong, and yet, for complicated reasons of pride, profit and prejudice, the hypothesis continues to be exploited by scientists, fund-raising enterprises, food companies and even governmental agencies. **The public is being deceived by the greatest health scam of the century."** George V. Mann, Sc.D., M.D., the co-director of the Framingham Heart Study emphatically states. Dr. Mann edited the book, *Coronary Heart Disease: The Dietary Sense and Nonsense.*

The majority of experts who created the new cholesterol guidelines in the U.S. have multiple financial ties to the pharmaceutical firms that make statin drugs, warn Ray Moynihan and Alan Cassels in their book *Selling Sickness*. One expert who developed the new lower cholesterol guidelines had financial ties to ten drug companies.

Other countries do not accept that research findings support a need to lower the guideline numbers for LDL cholesterol. Great Britain and European countries maintain guidelines of higher acceptable numbers, preventing an instant expansion of the market for statin drugs used to lower LDL cholesterol.

Danish physician Uffe Ravnskov, MD, PhD author of *The Cholesterol Myths,* describes how the emphasis on low LDL began with the landmark Framingham Heart Study. The study began following healthy people in the early 1950's to see who had a heart attack and if anything distinguished them from those who did not. High LDL cholesterol was only one risk factor of more than 240 risk factors.

"They have confused a statistical association with causation, " Dr. Ravnskov explains. "It is as if they saw a house burning and determined that the bigger the fire, the more firemen are present. And then concluded that firemen cause burning houses."

U.S. sales of statin drugs were more than $21 billion in 2006. Pfizer's statin, Lipitor, is currently the most prescribed medication in the world with most of its sales in the U.S. Marketing of statins is designed to lead people to assume lower LDL cholesterol will prevent heart attacks and death.

"Trials demonstrating a reduction in coronary artery disease from cholesterol lowering have not demonstrated a net reduction in all-cause mortality." States the ALLHAT study Web site. ALLHAT, completed in 2002, was the largest cholesterol study done in North America. The death rate was the same for both the group taking Lipitor and the group not taking any statin but making healthy life choices (no smoking, getting exercise, maintaining healthy weight). Drug sales people do not share those results with physicians.

"This is a perversion of science." John Abramson, M.D. of Harvard University spoke out against the change to LDL cholesterol guidelines. Making patients out of healthy people is a concern some doctors have with the new lower guidelines. Many noted physicians and researchers in the field of heart disease have warned that the scientific evidence does not support the lowered cholesterol guidelines. They are concerned the guidelines improve drug sales, not people's health.

No study has ever shown statins reduce death or heart attacks in women without heart disease, was the finding of a review of all statin trials by the University of British Columbia's Therapeutics Initiative (UBCTI) . And in more than one statin study the women in the statin groups had a higher overall death rate (all causes) than the women in the placebo groups. But drug companies use financial influence to ensure the deception to physicians and the public continues. Months after the UBCTI study results were published, the medical journal *Circulation* published an article in which 20 organizations endorsed cardiovascular disease prevention guidelines for women "preferably using a statin drug." Even though no clinical trial has ever shown statin drugs prevent heart disease in women or men.

The two most prescribed statin drugs, Lipitor and Zocor, are among the leading fifteen drugs for severe side effects in the FDA reporting system. In a Scandanavian Zocor study, more women died in the group taking Zocor than in the group taking a placebo (sugar pill). Merck doesn't tell doctors or the public those results.

> Hearing the truth about drugs is like hearing a mouse
> tap dance during a hurricane.

The UBCTI was created to help you hear the mouse. It provides physicians and the public with accurate pharmaceutical information. The UBCTI does not receive any financial support from drug companies. Their website is www.ti.ubc.ca.

How did the image of statins as wonder drugs to reduce heart attacks and death begin? One of the earliest statin trials was the Heart Protection study conducted in the U.K. with Zocor (simvastatin). The people in the study were high-risk patients who had coronary disease, occlusive arterial disease, or diabetes. The five-year study found a small reduction in coronary death rate of high-risk middle-age men. Drug companies latched onto that and created the image that lower LDL decreases the risk of death in any population. But the final sentence in the study's conclusion is: **"Benefit depends chiefly on such (*high risk*) individual's overall risk of major vascular events, rather than on their blood lipid concentration alone."** Blood lipid concentration is the same as serum cholesterol levels. Deceptive marketing by statin makers have misled doctors and the public alike for decades.

> There is more fiction in pharmaceutical marketing
> than in the film making industry.

"The ALLHAT was the only one of the 14 major statin trials included in the most recent review (*of studies*) not funded by a drug company. But the results showed that there was no less heart disease or death in the group with three times as many people taking statins." John Abramson, M.D., stated in an interview. Dr. Abramson, a cardiologist at Harvard Medical School co-authored with Dr. James Wright of the University of British Columbia, Vancouver, a review of all major statin trials. Their article, "Are Lipid-Lowering Guidelines Evidence-

Based?" was published in *The Lancet* in January 2007. "Our analysis suggests that lipid-lowering statins should not be prescribed for true primary prevention in women of any age or for men older than 69 years. High risk men (*those suffering from heart disease*) should be advised about 50 patients need to be treated for five years to prevent one event."

This means fifty men or more who already have heart disease must take statins daily for five years in order to prevent one less heart attack or stroke than a group of fifty men or more who have heart disease but are not taking any statin drug. In those five years there will still be heart attacks, strokes, and cardiac deaths in both groups, but there will be one less event in the statin group in those five years. This is the information that patients should be given before they make the choice to take a statin, rather than giving every patient the impression that by taking a statin drug and lowering their LDL cholesterol they are guaranteed less risk of a cardiovascular event.

Dr. Abramson explained in an interview, "Healthy lifestyle changes are a more effective, less expensive, and safer way to reduce your risk of heart disease and improve your chances of staying healthy overall."

"Stress has more deleterious effects on the heart than cholesterol." States Paul Rosch, MD a clinical professor of medicine at New York Medical College. He explained it is difficult to get research money if the research questions the role of cholesterol in heart disease. "Anyone who questions cholesterol usually finds his funding cut off."

Former NASA scientist, physician and astronaut Duane Graveline, MD, MPH, describes first-hand his serious side effect with Lipitor. Only on the statin for six weeks he went for a walk near his home and was found wandering, confused and reluctant to enter his own home because he did not recognize it. He was diagnosed with transient global amnesia

(TGA). Dr. Graveline came to his senses later that day. No one suspected the Lipitor. Six weeks later the TGA returned.

Dr. Graveline began to suspect Lipitor as the cause, although his physician told him that memory loss was not a listed side effect of the drug. Dr. Graveline contacted Joe and Terry Graedon, the husband and wife team who write the syndicated column, *"The People's Pharmacy."* Joe Graedon is a pharmacologist and Terry Graedon has a doctorate in medical anthropology and is a nutritional expert. Their website www.PeoplesPharmacy.com informs about drug side effects and natural remedies. With Dr. Graveline's permission the Graedon's published his letter in their column.

Hundreds of people responded to say they had also experienced severe memory loss while on Lipitor. Dr. Graveline did more research and found: "Thousands of cases of memory dysfunction have been reported to the FDA's Medwatch (adverse event) program but the agency still has not acted. And most practicing physicians are unaware of the problem."

Other statins are also linked to memory dysfunction, he explained. He has now learned the critical role that cholesterol has in the maintenance and healthy functioning of cell activity. On his website www.spacedoc.net he writes: "The same cholesterol that we have been led by the pharmaceutical industry to believe is public health enemy number one, is now proven to be absolutely vital in the formation and function of the trillions of synapses in our brains."

Think of the forming brains of children. In July 2008 the American Academy of Pediatrics (AAP) published a report recommending children as young as two-years-old have cholesterol levels tested and based on the arbitrary cholesterol guidelines established by the drug industry, children as young as eight-years-old could be put on statin drugs. The American Academy of Pediatrics receives millions of dollars from drug companies every year. Individuals within the AAP also have financial ties to the pharmaceutical industry. The Associated Press reported that Dr. Stephen Daniels, the lead author of the AAP report to test children's cholesterol and give children statin drugs, admits he has been a paid consultant to two drug companies.

The pharmaceutical companies that market statin drugs are trying to convince physicians and the public that tampering with children's and adult's serum LDL levels will prevent heart disease. Science does not support the drug industry's claims. According to the FDA Web site: "It is not clear why the lower levels of LDL cholesterol in the patients…did not lead to lesser amounts of plaque…" Plaque is the build-up on the walls of blood vessels that can cause serious health problems. Cholesterol has been falsely blamed for plaque formation by the drug industry. Seventy-five percent or more of your measurable cholesterol is made by your body, not from the food you eat. The body makes cholesterol for very important health reasons.

Both men and women with low cholesterol die earlier. The death rate was five times higher for elderly women with very low cholesterol in a French study reported in the medical journal *Lancet*.

A 15-year European study of 149,000 men and women aged 20 – 95 found low cholesterol was significantly associated with increased all-cause deaths.

Men with the lowest cholesterol died at younger ages of all causes, reports researchers at the University of Hawaii. The twenty-year study involved 3,500 Japanese-American men initially aged 51 to 73. At the end of the study the men were aged 71 to 93. Lead researcher Dr. Irwin Schatz warns, "Prudence dictates that we be less aggressive in lowering cholesterol in the elderly." The research findings also state, "Our data accords with previous findings of increased mortality in elderly people with low serum cholesterol, and show that long-term persistence of low cholesterol concentration actually increases risk of death. Thus, **the earlier that patients start to have lower cholesterol concentrations, the greater the risk of death."**

Marketing drowns out the truth of science. The number of young Americans taking statin drugs increased 68 percent between 2001 and 2006. There are approximately 4.2 million people (both men and women) aged 20 to 44 taking statin drugs. And now statin makers are pushing to have doctors prescribe their drugs to children.

The FDA-labeled side effects of statins (brand names: Crestor, Lescol, Lipitor, Mevacor, Pravachol, Vytorin, and Zocor) include acute liver disease, myopathy (neuromuscular disease), rhabdomyolysis (rapid breakdown of skeletal muscle tissue that can be fatal), pain in extremities and abdomen, angioedema (rapid swelling of the skin, mucosal and submucosal tissues), hepatitis, fatigue, headaches, influenza, sinusitis, upper respiratory tract infection, and serious problems during pregnancy and breast-feeding. Cholesterol is extremely important to the development of the fetus and is contained in breast milk. Pregnant women taking statins have a high rate of premature births and unhealthy babies. There is also the serious memory loss side effect that the FDA does not list.

There are more than 18 million Americans taking statin drugs as of January 2008. Statins have been on the U.S. market since 1987, yet in all that time the rate of deaths from heart attacks and strokes has not changed, according to the Center for Disease Control. Statins are not the wonder drugs preventing heart disease that the drug industry wants physicians and the pubic to believe.

The most recent attempt to bolster statin sales against the increasingly publicized scientific evidence against statin use is the Jupiter trial that was financed by drug maker AstraZeneca. The person in charge of the research, Dr. Paul M. Ridker, reports financial ties with more than ten pharmaceutical companies.

Although 90,000 people were screened, less than 18,000 people were selected to participate in the Jupiter study that took place in 1315 sites in 26 countries. "If you extrapolate that, it means there are not all that many people exactly like those who were studied," states Dr. Nieca Goldberg, director of the women's heart program at New York University Longone Medical Center. "My greatest concern is that there will be many people who don't fit the criteria of the study, but based on this they will get blood tests and statin therapy."

Although the drug-company-paid researchers are claiming a fifty percent reduction (in relative, not absolute terms) in the risk of serious heart problems in the statin group, in reality **all the people selected for the study were at low risk. The absolute difference in risk between the study's statin group and the placebo group was actually less than one percent.** The *New England Journal of Medicine* editorial about the Jupiter study concluded that treating 120 people who have the same health profile as the study participants for about two years would benefit only one person.

Besides this very slim indication of benefit, a serious fact of the study that is not widely publicized is that the statin group of people developed diabetes at a higher rate than the placebo group and as the study results report, the development of diabetes in people taking statin drugs has also been reported in previous statin trials.

The Jupiter study was stopped early, less than two years, ostensibly so that the placebo group could be offered statins for their presumed heart benefit. In other words, the negative outcome of the Jupiter study, the higher rate of diabetes that often leads to heart disease, that developed in the statin group was more significant than the positive outcome of one person out of 120 having a benefit over two years. But the study was stopped early because of the benefit? I suspect the drug company was fearful of how high the diabetes numbers would climb if the study continued.

How logical is it to take a drug that supposedly reduces risk of heart problems when the drug can in fact cause a disease that is a leading factor in heart disease?

Your body makes cholesterol for very good reasons. "Cholesterol is vital to proper neurological function. It plays a key role in the formation of memory and the uptake of hormones in the brain, including serotonin, the body's feel-good chemical...Cholesterol is the main organic molecule in the brain, constituting over half the dry weight of the cerebral cortex," (healthy adult brain is 85 percent water) explains Mary G. Enig, Ph.D. and Sally Fallon in their report, "Dangers of Statin Drugs: What You Haven't Been Told About Cholesterol-Lowering Medicines."

A reason for the high level of heart disease in western cultures may be an insufficient intake of B vitamins in relation to the amount of protein consumed. High levels of an amino acid called homocysteine were found to increase the risk of cardiovascular disease in a Harvard study of 15,000 physicians. Those physicians with the highest homocysteine levels had greater than three times the risk of suffering a cardiovascular event. Your body makes homocysteine when you eat animal and vegetable protein. Three B vitamins, B6, B12 and folate, help the body to safely break down homocysteine, reducing the level of this amino acid in your blood where high levels of it damage arterial walls.

You need to make healthy life choices of food, daily water intake and exercise so that your cholesterol levels are right for your body's needs. Statins create artificial cholesterol levels that are dictated by one-size-fits-all guidelines created by drug companies to sell drugs. It is about money, not health.

6

FRAUDULENT MARKETING CAUSES DEATHS

"If we put horse manure in a capsule, we could sell it to 95 percent of these doctors." Former CEO of Parke-Davis, Harry Loynd, told his sales people. Putting horse manure in capsules may be safer than many of the synthetic drugs that are now fraudulently marketed.

Legal drugs have something in common with illegal drugs, they are about vast amounts of money. The majority of people I met in the pharmaceutical industry are just looking for ways to get more of the wealth for themselves. All of the following drug industry practices are because the aim is money, not good health:

- Fraudulent marketing
- Marketing drugs to people who do not need to be medicated
- Inflating the price of drugs
- Inventing diseases from natural life processes
- Spending twice as much on marketing than on research
- Dishonesty about drugs' side effects
- Encouraging doctors to over-medicate
- Hiding negative clinical results from the FDA and other countries' licensing agencies

In September 2009, Pfizer pleaded guilty to a felony charge of fraudulent marketing of its anti-inflammatory drug Bextra. Pfizer will pay $2.3 billion in settlement.

There were 150 cases of alleged fraud by pharmaceutical companies on the Justice Department's docket as of February 2007. Selling legal drugs has something else in common with illegal drugs. The magnitude of money to be made entices many drug companies to step outside the laws.

Purdue Pharma agreed in May 2007 to pay $600 million in fines to the federal government and $19.5 million to 26 states and the District of Columbia for encouraging physicians to over-prescribe OxyContin, a strong, time-released narcotic. Three chief executives also agreed to pay $34.5 million in fines.

Purdue did market research with physicians in 1995 and learned doctors were concerned about the abuse potential of the drug. The company then gave false information to its sales force that OxyContin had less potential for addiction and abuse than competitor's products Percocet and Vicodin. The false claim became the leading point in an aggressive marketing campaign promoting OxyContin to physicians in general practice. These doctors have little training in treating chronic, serious pain or in recognizing signs of drug abuse. The marketing paid off for the drug company as OxyContin sales quickly grew to over a billion dollars a year. By the year 2000 some rural areas of the U.S. began to see soaring rates of addiction and the crimes that relate to drug use. In the Appalachian region OxyContin abuse is so common that it is called "hillbilly heroin." As I wrote this section of the book, I learned of a mother in Arkansas who had just buried her son who died from an overdose of OxyContin. He was only twenty-one years old.

In another case Bristol-Myers Squibb (BMS) pled guilty to federal criminal charges by the Department of Justice. BMS's CEO resigned unexpectedly in September 2006. F.B.I. agents searched the company's New York City Park Avenue headquarters investigating charges of a secret agreement BMS had made with Canadian drug maker Apotex. The Canadian firm has an FDA approved generic blood anti-clotting drug that cut into world sales of BMS's Plavix the last quarter of 2006. Worldwide sales of Plavix in 2005 were $5.9 billion. BMS pled guilty to offering bribes to Apotex to stop production of their generic blood anti-clotting drug. BMS was ordered to pay a paltry fine of $1 million.

Plavix is a once-a-day tablet. The cash price (no insurance) of 30 tablets of Plavix costs approximately $128 to $155 in U.S. drug stores. The price of the Canadian generic drug, Clopidogrel, for 100 tablets purchased online from a Canadian pharmacy is $87.

BMS also paid $515 million in September 2007 to resolve U.S. Department of Justice allegations of illegal drug marketing and pricing. Based on the evidence of several whistleblowers, the government charged BMS illegally promoted Abilify, an antipsychotic drug, to physicians for treatment of children and for dementia in the elderly. Abilify is not FDA approved for children and the FDA had required a Black Box Warning against its use for dementia-related psychosis.

According to federal investigators, BMS assigned its sales people to call on pediatricians (children specialists) and nursing homes to fraudulently market Abilify from 2002 through 2005.

The government also charged BMS with defrauding Medicaid by fraudulent and inflated prices on several oncology drugs and the antidepressant drug Serzone, a drug that has high risk of causing liver damage and liver failure.

Further charges against BMS involved illegal inducements to physicians and other healthcare providers to purchase BMS drugs. The Department of Justices alleges BMS paid physicians in the form of excessive consulting fees and paid for travel to luxurious resorts.

Whistleblower Dr. Susan Molchan, a former clinical researcher for the National Institute of Health (NIH), testified before Congress in June 2006 that her supervisor at NIH made a secret deal with Pfizer. Dr. Molchan testified that a collection of spinal fluid samples disappeared from her laboratory freezer at NIH. The samples were painfully given by Alzheimer patients for NIH research and were extremely valuable. The procurement costs alone for the tissue samples cost taxpayers $6.4 million.

Dr. Molchan's testimony and other data gathered by Congressional investigators led to felony charges brought against her supervisor, Dr. Trey Sunderland. Investigators uncovered Pfizer payments to Dr. Sunderland of more than $600,000. No charges were brought against Pfizer. The drug company now has the leading drug for Alzheimers. Aricept sales in 2006 were $1.4 billion.

In another case Pfizer pled guilty to two felonies and paid $430 million in penalties to settle federal criminal and civil charges in May 2004. Pfizer assigned their sales force to fraudulently promote the epilepsy drug Neurontin for off-label use. The company turned Neurontin into a blockbuster drug with tactics like paying doctors to listen to pitches for unapproved uses and paying for physicians to take luxury trips, federal prosecutors said. As much as 90 percent of Neurontin sales were for off-label uses according to some estimates. The fraudulent marketing went on for several years, taking sales from $97.5 million in 1995 to nearly $2.7 billion in 2003.

A former TAP Pharmaceutical Products Inc. vice president of sales blew the whistle on the drug company's kickbacks to physicians to promote Lupron, a drug for advanced prostrate cancer. Douglas Durand alleged that TAP sales representatives were directed to provide physicians with free samples of Lupron and encouraged the doctors to profit from the gifts by billing Medicare and Medicaid at $500 per dose.

In October 2001 TAP agreed to pay $875 million to resolve criminal charges and civil liabilities for fraudulent drug pricing and marketing of Lupron. TAP agreed to pay the federal government for filing fraudulent claims with Medicare and Medicaid. The drug company also paid the fifty states and District of Columbia for filing fraudulent claims.

Schering-Plough Corporation pled guilty and agreed to pay $435 million to settle conspiracy charges in August 2006. According to the U.S. Justice Department, Schering-Plough engaged in illegal sales and marketing for several chemotherapy drugs. It promoted the drugs for uses not approved by the FDA. The drug company was also cited for using an illegal kickback scheme to encourage physicians to use their drugs. The case was the third multi-million dollar government settlement by the company in five years.

Sanofi-Aventis Pharmaceuticals paid $190 million to the federal government and several states in September 2007 to resolve allegations of Medicare fraud. The government alleged that the drug company set fraudulent and inflated prices for the drug Anzemet, used to prevent nausea and vomiting due to chemotherapy and radiation treatments.

As of July 2008 ten states have filed Medicaid fraud lawsuits against Eli Lilly for the antipsychotic drug Zyprexa, FDA approved for the treatment of schizophrenia and bipolar disorder in adults. Twenty more states are considering lawsuits. **Zyprexa is one of the top fifteen most deadly drugs** according to a September 2007 report.

Montana Attorney General Mike McGrath said, "Lilly allegedly gave kickbacks to doctors and improperly promoted the drug to nursing homes as a sedative." The Montana suit contends that Lilly "instructed its representatives to minimize and misrepresent the dangers of Zyprexa, affirmatively and consciously placing company profits above the public safety.

This failure to warn was designed and intended to maximize company profits."

Zyprexa use in the elderly, those 65 years and older, may be especially dangerous. A new study of the FDA adverse events reporting system found that the elderly accounted for one-third of the serious adverse events reported although the elderly represent only 12.6 percent of the total population. Between 1998 and 2005 more than 1000 deaths were reported to the FDA for Zyprexa. The FDA admits reported numbers reflect only one to ten percent of the actual. That means the actual number of deaths from Zyprexa between 1998 and 2005 was in the range of 10,000 to 100,000.

Zyprexa is one of three antipsychotic drugs in the top 15 most deadly drugs, according to the FDA reporting system. Risperdal (also known as Ridal, Rispolept, Rispen, and Belivon) and Clozaril (also known as Leponex, Fazacio, and in Canada Gen-Clozapine) are the other two. Between 1998 and 2005 there were 5,375 deaths reported to the FDA for these three antipsychotic drugs. Therefore the actual number of deaths can be estimated at between 53,750 and 537,500 in seven years. Those are just deaths in the U.S. The worldwide numbers are not known.

Mississippi is another state suing Eli Lilly for fraudulent marketing of Zyprexa. The lawsuit claims Lilly trained and instructed its sales force to attempt to expand the drug's market by convincing primary care physicians to prescribe Zyprexa for mood, thought and behavioral disturbances. Lilly, the suit alleges, provided its sales force with marketing brochures with hypothetical patient profiles that included "patients complaining of symptoms such as anxiousness, irritability, mood swings and disturbed sleep" encouraging doctors to prescribe Zyprexa for patients with these symptoms. The lawsuit states, "As a result Mississippi is spending millions of dollars on Zyprexa for patients who are not indicated for the drug, and further, who are being harmed by it."

Global sales of Zyprexa in 2006 were $4.36 billion. In September 2003 the FDA required warnings for increased risk of diabetes with Zyprexa. The warning came a year behind other countries' realization that Zyprexa use was linked to diabetes. In 2002 British and Japanese pharmaceutical regulatory agencies warned physicians about the increased risk of diabetes and hyperglycemia with Zyprexa. Diabetes is a serious condition leading to many other health problems including higher risk of heart attacks and strokes. Britain and Japan required that special warnings for the risk of diabetes be added to Zyprexa packaging in their countries. The majority of Zyprexa sales are in the U.S.

Hundreds of millions of dollars of states' Medicaid money have been spent and continues to be spent for Zyprexa, one of the most deadly drugs. Billions of Medicaid dollars will be required to treat the diabetes and the serious health conditions people with diabetes suffer. More costs to flow downhill to you the taxpayer.

The increase in diabetes does not worry Eli Lilly executives. The drug company has popular drugs to treat diabetes. Your tax dollars will continue to flow into Lilly's coffers.

Zyprexa has caught the attention of the U.S. Congress. Senator Charles Grassley (R-IA), a member of the Senate Committee on Finance, sent a letter to Lilly Chairman and CEO Sidney Taurel on April 4, 2007. Senator Grassley wrote, "I have an obligation to ensure that the public's money is properly spent to provide safe and effective treatments to the vulnerable populations that are beneficiaries of the Medicare and Medicaid programs." Senator Grassley requested all documents about Zyprexa.

The New York Times was provided with internal Lilly documents showing the original clinical study results of Zyprexa found three and one-half times higher risk of hyperglycemia (high blood sugar) than the people in the trial taking the placebo. Hyperglycemia is a precursor to diabetes. Another Lilly report shows that after examining 70 clinical trials for Zyprexa, 16 percent of patients taking Zyprexa for one year gained more than 66 pounds. Obesity is a precursor to diabetes. The company had not disclosed any of these findings to the FDA.

By 1999 physicians had become concerned about the considerable health issues that the weight gain and hyperglycemia caused for patients taking Zyprexa. According to internal documents Lilly trained their sales people to mislead physicians about the risk of diabetes associated with the drug and to illegally promote Zyprexa for off-label use with children and elderly. Zyprexa is only approved by the FDA for schizophrenia and bipolar disorder in adults. It is not approved for children for any use.

Ellen Liversridge's son gained 100 pounds when he took Zyprexa. He died four days after going into a coma in October 2002. His death at age 39 was due to "profound hyperglycemia."

"We must embrace the fact that many physicians are curtailing their use of Zyprexa particularly in the moderately-ill patient and in the maintenance phase," states an internal Lilly memo titled "Diabetes Update" from July 7, 2003. The memo goes on to describe the marketing strategy to be used. Drug sales representatives were to inform doctors Lilly would indemnify them. It means Lilly would cover any lawsuits filed against the physician as a result of Zyprexa. Lilly had used the same marketing strategy after the risk of suicide from Prozac became public. The memo claims, "Our experience with Prozac confirms the impact and goodwill of such an initiative." In plain words, Lilly had no concern for the life and health of anyone taking their drugs. Lilly's only concern was there would not be any drop in prescriptions for their drugs.

Dr. David Graham, an FDA scientist and whistleblower, testified on February 13, 2007 to a Congressional subcommittee. He told them that the FDA knew "for a long time" about the risk of weight gain and diabetes with Zyprexa. He recommended Congress investigate the FDA's handling of the important health issues with Zyprexa.

Zyprexa increases the risk of obesity, hyperglycemia, diabetes, hypertension, heart attacks and stroke. Any of these health issues can shorten a person's life. Alex Berenson, investigative reporter, warned in *The New York Times* of the great cost to the public for these health conditions. "Mental illness is itself a money sponge, an expense born largely by tax dollars. But that cost may be dwarfed by the bill to manage the heart attacks and amputations that diabetes bestows."

And the drugs used to treat diabetes cannot be trusted as safe. According to Dr. David Graham, Avandia has caused more than 200,000 people to suffer heart attacks or strokes since it was FDA-approved to treat diabetes in 1999. On July 30, 2007 the FDA convened a panel of experts to address the high risk of heart failure from Avandia made by GlaxoSmithKline (GSK). Dr. David Graham encouraged the FDA to remove the drug from the market. He estimates that as many as 2,200 people will suffer a serious cardiovascular event for every month Avandia remains on the market. The FDA panel of experts, with their financial ties to the drug industry, voted to leave Avandia on the market.

It Is Only Money

Other than the fines to the three Purdue Pharma executives who were fined in the OxyContin case, there have not been any reports of fines or prison sentences for any drug company executives.

There are U.S. business laws that do not permit a person to commit a crime and then hide within the entity of a corporation. By law, incorporated companies are an entity. But the intent of the law is not to allow people to commit crimes and not be held responsible by hiding behind the veil of a corporation. It is people within the drug companies who are making decisions to commit fraud that causes injury and death. There has truly been deceit with intent to harm by some drug companies. Why has no attorney general attempted to, as they say in law, "pierce the corporate veil"?

Martha Stewart went to prison for lying to the government. Her lies did not harm or kill anyone. The former director of China's State Food and Drug Administration was sentenced to death on May 29, 2007. He was convicted of taking bribes from drug companies and dereliction of duty. Perhaps we would see improvement in the drug industry if people were held accountable for their management decisions and not allowed to hide behind the corporate veil.

What is the Aim?

No one in the drug industry is looking for methods to make medications safe and affordable for patients who truly need them. The industry is caught up in a vicious cycle. Pharmaceutical firms have become some of the wealthiest companies in the world. Their senior management teams' attention is always focused on financial numbers. Continuously striving to have greater bottom line numbers every quarter and every year. Stock values are expected to always be on an upward trend. Attention is on money, not on medications that are safe and are truly about health.

The pharmaceutical industry's disregard for customers reminds me of Enron. In his book about Enron, *Conspiracy of Fools*, Kurt Eichenwald writes, "It set off what became a cascading collapse in public confidence, sealing the final days of an era of giddy markets and seemingly painless, risk-free wealth."

Is it cost effective both in terms of the price of medications and in terms of physicians' time to continue in the old pattern of door-to-door selling? The amount of marketing fraud that has been discovered recently, which can only be a small percentage of the actual, is serious enough to question the validity of drug sales people calling on physicians. In today's high tech environment with the Internet and medical journals in print and on-line, physicians have easy access to pharmaceutical information and new scientific findings (if drug firms haven't repressed negative research results.) Perhaps it is time for a paradigm shift. This is the twenty-first century. Why are drugs still being peddled door-to-door by legal drug pushers?

7

REPORTING SERIOUS SIDE EFFECTS, OFFICIAL NUMBERS ARE MISLEADING

November 14, 2005

I arrived at Dr. S's office just as they were about to close for lunch. Dr. S. graciously came out to sign for samples. "I had a patient on product C hospitalized for rhabdomyolysis." My heart sank. Rhabdomyolysis is a debilitating muscle-wasting condition that can be fatal and is the most serious side effect of the class of drugs called statins. I told Dr. S. I would need some patient information in order to make a formal report to the corporate office. Pharmaceutical companies must report adverse events to the FDA. Initially Dr. S. was hesitant. "Well the patient is okay now and besides he is a former alcoholic." Alcoholism increases the risk of rhabdomyolysis because alcoholism damages the liver. Dr. S. knew he would get calls for verification of information and he did not want to be bothered. I pressed him by explaining that I had a legal obligation to report the adverse event (serious side effect) the same day.

When a physician tells a drug representative about any type of adverse events a patient has from a medication, a report must be made by the drug rep. to the drug company corporate office the same day. The drug company is legally required to report adverse events to the FDA. This is how the FDA keeps statistics about drugs. The reporting of numerous adverse events, especially life threatening events, may lead to requiring additional warnings be added to the drug's package insert. The FDA can require a drug be withdrawn from the market, but that rarely occurs. Withdrawal of drugs usually occurs voluntarily when lawsuit settlements became too large for drug companies to ignore, but that may take several years.

There is no mandatory reporting required by physicians in the U.S. therefore only a small percentage of adverse events are actually reported to the FDA. Sometimes the patient just stops taking the medication and doesn't inform the physician. Often physicians don't say anything to drug salespeople because physicians know it will require their time or they are afraid of lawsuits. Patients can report adverse events directly to the FDA. Directions are provided at the end of this chapter.

November 15, 2005

The messenger must be shot. "Kay, what is this about a case of rhabdomyolysis from product C?" The computerized report I had completed about Dr. S's patient had been circulated to four managers and my team members. Now I had to defend my actions. One of the other sales representatives on the team called me to accuse me of "misdiagnosing." I had to remind her Dr. S., a respected medical doctor, had diagnosed and treated the patient. I simply fulfilled my legal obligation to report an adverse event.

They did not want to hear the bad news because of the seriousness of rhabdomyolysis. Bayer's Baycol had been withdrawn from the market after hundreds of cases of the potentially fatal condition were reported. My team was ready to blame me for the fact that a report of rhabdomyolysis was going to be linked to product C. After all, if the news got out it could prevent growth of market share. That would not win them any bonuses. I felt both a legal and an ethical obligation to report the case even though Dr. S. had been hesitant. Given the reactions of others in the work group, I have sincere doubts if the report would have been made if a different salesperson had spoken with Dr. S.

Adverse events need to be reported. Reports of adverse events to the FDA after a drug has been approved are extremely important because it provides the data for the FDA to consider if a drug should have additional warnings or is too dangerous compared to its health benefits and should be withdrawn from the market. As the Vioxx tragedy described in the next chapter demonstrates, serious adverse events suffered by patients during clinical trials can be buried by vigorous, expensive marketing campaigns.

Drugs Withdrawn From the Market

Ten drugs were withdrawn from the U.S. market from January 1, 1997 through December 2000 according to a report from the U.S. General Accounting Office (GAO) to senators Tom Harkin, Barbara Mikulski, and Olympia Snowe. The GAO looked only at prescription pharmaceuticals, not vaccines or over-the-counter medicines. The GAO letter explains that withdrawal of drugs does not provide a complete picture of drug safety.

> "Drug withdrawals do not reflect a judgment concerning the absolute safety of a drug but reflect a judgment about the risks and rewards of a drug in the context of alternative treatments."

According to the FDA, a drug that has had documented deaths may remain on the market until a new, safer drug with similar benefits is available to replace it. This was the case for Rezulin, a medication for diabetics. It caused liver failure in some people but remained on the market from January 29, 1997 to March 21, 2000. It was withdrawn from the market when a safer drug became available. A drug may be withdrawn on the basis of relatively few adverse events reports if other medications that provide similar benefits and higher safety records is available. As was the case of the antibiotic Raxar, approved by the FDA on November 6, 1997 but withdrawn on November 1, 1999 after cases of potentially fatal cardiac arrhythmia were reported.

However there are cases when dangerous, deadly drugs like Vioxx and Zyprexa remain on the market even though safer, effective medications are available.

Are ties between the pharmaceutical industry and the government agencies that are supposed to regulate it creating risks to public safety? There are definitely grounds for concern. A survey taken in 2006 by the Union of Concerned Scientists found one in five FDA scientists had been asked to manipulate or exclude data from research studies. In some FDA departments as many as 75 percent of the scientists were "not at all" or "only somewhat" confident that the "FDA adequately monitors the safety of products once they are on the market." The survey also found 61 percent of the scientists responding knew of cases when political appointees at the FDA had inappropriately interjected themselves for the protection of pharmaceutical companies.

Scientists and physicians outside of drug companies have been sued for finding the truth. When researchers concluded the anti-AIDS drug, Remune, provided no real benefit to patients, the scientists were sued for $10 million by the drug manufacturer for damaging business. In another case a scientist's contract with a drug company was terminated when her research revealed a serious side effect of the drug Deferiprone, used for a blood disorder.

Drug companies can be extremely successful at hiding the truth. In the early 1990's Betty Dong, a pharmacologist at the University of California at San Francisco, found that a generic thyroid hormone was just as effective as the brand drug

Synthroid made by Knoll Pharmaceuticals. The generic drug was much less costly for patients to purchase. Knoll Pharmaceuticals had funded the research at UCSF and for seven years they successfully blocked publication of Dr. Dong's findings. The fraud was finally discovered in 1997 and Knoll had to pay 37 states $42 million for consumer fraud.

Twelve drugs have been withdrawn from the market in the seven years between 2000 through March 30, 2007. The most recent is Zelnorm (tegaserod maleate). A drug given to women for what some experts consider the drug industry contrived "irritable bowel syndrome" and chronic constipation. Zelnorm was withdrawn on March 30, 2007 because of severe cardiovascular events including heart attacks, strokes, and angina. The drug was on the market nearly five years. Do not trust that a drug is safe from life-threatening side effects just because it has been on the market for years. Vioxx was on the market for five years before being voluntarily withdrawn.

The vast majority of physicians do want to help people. They are limited by the drug-oriented medical education and by a medical environment that encourages prescription drug therapy. As more patients request suggestions for life style modifications for their health issues rather than being put on drugs, we will see more physicians practicing integrative medicine. The most effective way to decrease the number of deaths each year due to drug adverse events is to decrease the number of people taking prescription drugs. Life style changes improve health by treating actual causes of health conditions. The vast majority of drugs only treat symptoms. They don't heal.

Women Beware

Eight of the ten drugs withdrawn from the market in the four years between January 1997 and December 2000 posed greater health risks for women than for men. Four of these drugs were being prescribed more frequently to women. The other four drugs withdrawn that posed a greater health risk for women were prescribed to both women and men. The FDA has not identified specifically why these four drugs are a greater health hazard to women. Physiological differences between women and men may put women at greater risk with some drugs.

Be cautious about taking prescription drugs. Do not trust a drug is safe for you because your friend takes it without side effects. Your body, family history and other health factors are unique. Take steps to protect your health. Read the package insert, especially the warnings, cautions, and adverse events in clinical trials. Ask your pharmacist or physician if a class of drug has greater health risks for women. Get your family computer nerd to investigate on-line. There are now websites for people to post side effects they experience. Two such sites are http://www.askapatient.com and ttp://www.dangerousmedicine .com.

Do not assume a prescription written by your trusted physician is harmless to you. There are more than 5,700 prescription drugs listed on the FDA website. That number can fluctuate slightly as drugs are removed or added. It is a Herculean task for physicians to know all details of every product. Drug representatives may not tell physicians the negative facts about drugs. In fact the sales force often is not aware of additional risks to women because the drug companies may not be informing them.

Women are twice as likely as men to die in the first year after a heart attack. If a drug has increased risk for heart problems (shown on the package insert as cardiovascular risks) as a possible adverse event women should be especially concerned about using the drug. Half of women suffering the first heart attack die within a year. The death rate for men having their first heart attack is 25 percent. Read the package insert for every drug you are prescribed. Consider what benefits you can expect from a pharmaceutical drug and weigh that against the risks. And look for non-drug therapies. Physicians prescribe drugs because drug therapy is a cornerstone of their medical education.

Ethnicity and Age Can Alter the Power of Drugs

Some drugs have a stronger affect on various ethnic groups. For example the statin class of drugs is twice as powerful in people with Asiatic heritage. The recommended starting dose of this class of drugs is one-half of the recommended starting dose for people of other ethnic groups.

The popular ulcer drug Prilosec accumulates in the blood of Asians at four times the rate it accumulates in other ethnic groups. Studies have also found that Prilosec accumulates at a much higher rate in the blood of elderly people of all ethnic groups, and stays in the bloodstream of elderly people 50 percent longer.

The package insert should describe any ethnic or age variation of a drug. The information may be buried in the "Clinical Pharmacology" section. Read through the package insert yourself before taking a drug for the first time. **You can also call the FDA with any questions about a drug. The toll free number is 1-888-463-6332.**

Always verify a prescription drug has been approved for children or teen-agers. Most drugs are not tested on children and young people under the age of 21 and therefore are not FDA approved for children. In their hyper-aggressive marketing some drug sales people mislead physicians by improperly marketing drugs (referred to as off-label) for use in children when in fact the drugs have only been clinically tested on adults. Children are at a much higher risk of suffering serious side effects from any drug.

Recently the Texas Attorney General brought a case against Janssen, a Johnson and Johnson drug subsidiary, for improperly marketing a powerful psychiatric drug for use in children. The drug had never been clinically tested on children or approved for use in children by the FDA. Janssen's off-label marketing practices cost the Texas Medicaid program $117 million over five years, the state's assistant attorney general told Congress in February 2007.

Only you can protect your child from potentially harmful drugs. Check with your pharmacist or call the FDA. Never assume because a physician has provided the prescription the drug is safe for children. As the Janssen case demonstrates, physicians may be misinformed by aggressive drug salespeople who are trying to earn large bonuses.

Drug Interactions Can Be Lethal

When taking prescription medications more is not necessarily better for your body. One of the reasons that higher rates of adverse events occur after a drug is approved and becomes available to the public is because clinical research can only be reasonably done on a few thousand people. When a drug is approved and the number of people taking it grows to hundreds of thousands or even millions, the opportunity for drug-to-drug interactions becomes a concern. With the larger number of people taking a drug comes greater variation in the circumstances of use. That is why adverse events (side effects) may not show up in clinical trials but will become evident after FDA approval.

Since there are more than 5,700 prescription drugs it is not feasible to include tests for drug interactions with all other medications during the clinical trials required before a drug receives FDA approval. The chance of drug interactions increases when drugs are metabolized in the same areas or pathways of the body. Interactions that cause adverse events can also occur between foods and pharmaceuticals. There is no guaranteed safety with any drug because of the infinite number of variables. You know your body better than your physician. Watch for changes that may occur when taking any drugs. Even something such as constipation should be reported as an adverse event. Lotronex (alosetron) was withdrawn from the market in the year 2000 because of severe constipation and ischaemic colitis. Both can be fatal and were for some patients taking Lotronex.

Reporting an Adverse Event Directly to the FDA

The Food and Drug Administration (FDA) has an on-line reporting form for any person who has suffered a "serious drug adverse event, potential and actual product use errors, or product quality problems associated with the use of FDA-regulated drugs".

> Go to www.fda.gov./medwatch
> Click "How to Report"
> In the third paragraph click "Online Reporting Form"

At the bottom of the page click "Begin" to begin the Medwatch Online Reporting Form. You will receive an email message confirming the receipt of your report. **You can also call the FDA to report an adverse event. The toll free number is: 1-800-332-1088.**

Adverse events that involve vaccines should be reported to the:

Vaccine Adverse Events Reporting System (VAERS) on-line at: https://secure.vaers.org/VaersDataEntryintro.htm

As mentioned earlier, you can call the FDA directly with questions about a drug. The telephone number is: 1-888-INFO FDA (1-888-463-6332).

It is important the public report adverse events from prescription drugs because physicians rarely do. The FDA's lack of action even when thousands of adverse events have been reported for a drug does not inspire physicians to make the effort. But reported adverse events become public record and can be extremely useful to independent, nonprofit organizations that monitor the pharmaceutical industry. One of these is the Center for Science in the Public Interest (CSPI), a nonprofit organization based in Washington, D.C. CSPI continuously gathers safety facts about prescription drugs and alerts the public and advises Congress about dangerous prescription drugs. CSPI and other nonprofit organizations have taken over the watchdog job that the public expects from the FDA.

8

POLITICS OF DANGEROUS DRUGS

Vioxx killed more Americans than the Vietnam War. Between 89,000 to 139,000 U.S. deaths and worldwide deaths are estimated to be 150,000 to 200,000. While people were dying, Merck salespeople and managers were collecting hefty bonuses.

How do these dangerous drugs get approved? It begins with ignoring the obvious in study results and hiding or minimizing any negative data. Merck received FDA approval for Vioxx, a pain reliever in the class known as COX-2 inhibitors, in May 1999. Results of one study published in the *New England Journal of Medicine* (NEJM) in November 2000 reported five times as many heart attacks and twice as many cardiovascular problems (such as stroke) in the group of patients taking Vioxx. The other group of patients in the clinical trial was given naproxen, a non-steroidal anti-inflammatory painkiller. Merck researchers tried to explain the difference in heart attacks and cardiovascular problems by suggesting the people in the group taking naproxen may have had some heart protective attributes similar to low-dose aspirin. Apparently the FDA bought it. However not everyone did.

After the NEJM article, several cardiovascular (heart) researchers associated with the world-renowned Cleveland Clinic reviewed the data from four studies involving COX-2 inhibitors. They became so concerned they published a Special Communication in the *Journal of the American Medical Association* (JAMA) in August 2001 warning that there was risk of COX-2 inhibitor drugs causing cardiovascular events. These researchers called for additional trials to "characterize and determine the magnitude of the risk."

Drug companies finance the majority of clinical trials. Merck wasn't going to design a study to look for serious adverse events of Vioxx. Merck had put substantial money into marketing Vioxx and it paid off when U.S. doctors wrote 5 million prescriptions in the first seven months Vioxx was available. To counter the Cleveland Clinic publication in JAMA, Merck designed drug trials and published results that favored Vioxx and excluded the cardiovascular events. Drug studies can be designed to find the desired outcomes and hide the negative outcomes. It is not good science, but goes on in the drug industry all the time.

"Don't bring up the heart risks" the 3000 Vioxx salespeople were told in a February 9, 2001 internal memorandum. Merck aggressively marketed Vioxx, creating sales campaigns call Project XXceleration and Project Offense. The training of the Vioxx salespeople was designed to treat physicians' safety concerns as obstacles to sales that had to be overcome. Generous bonuses were given to the Vioxx salespeople and managers as people taking Vioxx were dying.

"Simply incomprehensible." Stated an FDA warning letter to Merck. The object of the letter was a May 2001 press release from Merck. The headline read, "Merck Confirms Favorable Cardiovascular Safety Profile of Vioxx". An outright lie was released to the public via the news media the public trusts. Merck's marketing was so aggressive and inaccurate that in 2001 the FDA required the drug company add a warning of possible cardiovascular events to the Vioxx label and cited the company for minimizing cardiovascular risk in Vioxx promotions.

What was the FDA doing besides writing letters? A fundamental principle of administrative law is that once a government agency approves a product or gives a license, the burden of showing good cause to take away the right to market a product is on the agency. The FDA had to prove that Vioxx was doing more harm than good and most of the research the agency had to rely on to make the decision was from Merck.

There was also behind the scenes political pressure. All the large pharmaceutical companies, including Merck, had given millions of dollars to the majority party and presidential nominees' campaigns. The senior people of the FDA are political appointees. It is not easy to be a person owing allegiance to a group in order to hold onto one's job and also have pressure from other groups to remove from the market a product that is making billions of dollars, a percentage of which would flow back into the FDA. It is an awkward system to say the least. People see what they want to see. FDA leaders did not want to "see" that thousands of Americans were dying because of Vioxx.

Finally on September 30, 2004, a full 3 years after independent researchers from the Cleveland Clinic published their warning in JAMA and 5 years after Vioxx received FDA approval, Merck voluntarily withdrew Vioxx from the market. The company was under public pressure and threatened by lawsuits due to the number of cardiovascular deaths associated with Vioxx.

Within days of the withdrawal of Vioxx the *Wall Street Journal* uncovered an analysis done by Dr. David Graham, an FDA safety official. The study examined data from 1.4 million Kaiser Permanente members and concluded 27,000 heart attacks and sudden cardiac deaths of Kaiser members may have been prevented if the patients had not taken Vioxx. This important study that was not sponsored by Merck and was completed within the government agency meant to guard the public against deadly drugs was never allowed to be published while Vioxx was on the market.

There were serious internal struggles at the FDA. The Associated Press obtained internal FDA emails warning Merck executives prior to the drug's withdrawal that an FDA safety official was about to go public with the life threatening dangers of Vioxx. One FDA official went so far as to advise Merck to "issue an official rebuttal" to Dr. Graham's analysis that Vioxx increased the risk of cardiovascular events.

FDA Gave Vioxx a Vote of Confidence

Shockingly, an FDA advisory committee gave Vioxx and the two other COX-2 inhibitors a vote of confidence. In February 2005 the advisory committee met to discuss the cardiovascular risks posed by COX-2 inhibitors. Because Vioxx had been voluntarily withdrawn by Merck it was considered by the committee along with two Pfizer drugs in the same class, Bextra and Celebrex. The 32 scientific experts chosen by the FDA to evaluate the drugs voted to keep all three on the market with heightened warnings listed on the package inserts. What this means is Merck could remarket Vioxx if they chose to. The FDA never lost Merck's goodwill. Only the public lost.

The *New York Times* requested the Center for Science in the Public Interest (CSPI) to verify any affiliations the members of the advisory committee had with drug companies. The CSPI research uncovered direct affiliations between ten of the committee members and Merck, Pfizer and Novartis. Novartis was developing a COX-2 inhibitor at the time. The Federal Advisory Committee Act prohibits scientists with direct conflicts of interest from serving on panels offering advice to federal regulatory agencies. The presence of ten scientists with direct ties to the companies with the drugs being considered appears to violate the law. The FDA showed no concern. Based on this information the *New York Times* analyzed the votes of the committee members and found the advisory committee would have voted against Vioxx and Bextra remaining on the market if the scientists with conflicts of interest had not been on the committee.

The Center for Science in the Public Interest also did an analysis of the 237 COX-2 inhibitor studies published (that excluded Dr. Graham's study because it was never published) after Cleveland Clinic's August 2001 JAMA article requesting studies be done to measure the risk of heart attacks, strokes and other cardiovascular events from COX-2 inhibitors. CSPI found no researchers had conducted such a study, not even those funded by public money rather than the drug companies. The vast majority of the studies were looking for additional uses for the drugs. When an off-label use of a drug can be shown effective by clinical research, the drug company can seek FDA approval for the new use. If the use is approved it creates a new on-label use for the drug, opening additional markets and generating potentially additional millions or even billions of dollars for the company. It may also place millions of additional people at risk of serious adverse events.

Merck finally agreed to pay a $4.85 billion settlement. On November 8, 2007, more than three years after Vioxx was withdrawn from the U.S. market, Merck agreed to the largest settlement in history. The drug company faced more than 47,000 individual personal injury lawsuits and 265 potential class action cases filed by people or family members worldwide that claimed injury or death due to Vioxx. Merck continues to stress that it is not admitting any fault.

"Immediately cease the dissemination of promotional materials for Celebrex and Bextra." The FDA letter to Pfizer stated. Pfizer had taken full advantage of the removal of Vioxx from the market by waging a ferocious and deceptive advertising campaign that included television ads, print ads, direct mail brochures and a 27 minute "infomercial". The marketing from all these sources was so misleading even the FDA was driven to action, sending a strongly worded letter to Pfizer to "immediately cease the dissemination of promotional materials for Celebrex and Bextra." Unfortunately the content of FDA letters are not provided to the public on national television and the Internet. The letter stated the ads were misleading because they omitted safety warnings, made unproven effectiveness claims and implied that the drugs were treatments for arthritis. The FDA does not have any authority to make drug companies pay a penalty for misleading advertising. Do you think drug companies are really threatened by warning letters?

Pfizer voluntarily withdrew Bextra from the market in January 2007 because of mounting lawsuits due to the risk of cardiovascular events and serious, potentially life-threatening skin reactions called Stevens-Johnson syndrome and toxic epidermal necrolysis.

Celebrex remains on the market but the FDA required the addition of a Black Box Warning about increased cardiovascular and gastrointestinal risks. A Black Box Warning is the strongest warning that can be issued. It shows up on the package insert (PI) with a heavy black box around the warning, drawing immediate attention to anyone looking at the PI of a drug. If you find a Black Box Warning on the PI of any medication you or a loved one uses, take the warning very seriously. The next step the FDA will take after requiring a Black Box Warning is to require the withdrawal of the drug from the market. That is if the FDA will listen to its own non-biased safety experts.

9

MEDICATING THE HEALTHY

Healthcare spending is on a path similar to a Space Shuttle launch – up, up, and away. Many companies you would not normally consider in the medical field are jumping on as a ride to riches. Since the mid-1990s drug firms have been increasing the outsourcing of drug clinical trials. A whole new industry of contract research organizations (CROs) has quickly flourished. Contract companies of any type remain in business only if they continue to renew or gain new contracts by pleasing their customers. CROs have a vested interest in providing drug firms with clinical trial data that will get a new drug approved, or gain new on-label uses for drugs already on the market. As we saw with the COX-2 inhibitor studies, if a research study is not set up to test a specific hypothesis, such as the risk of cardiovascular events, it may not be thoroughly documented in the research results. CROs are not in business to find bad news. Their objective is to get drugs approved.

The immense profit of selling drugs has created strange new partnerships. Some CROs are owned or partly owned by advertising companies. Donald L. Barlett and James B. Steele in their book *Critical Condition How Healthcare in America Became Big Business and Bad Medicine* give the perspective of Thomas L. Harrison, CEO of a large marketing and advertising firm, "With the goal today being faster, smarter and more efficient commercialization of new drugs, we felt that we needed to get closer to the test tube. To actually work with clinical scientists to develop drugs with a keen eye toward the needs of the market place that will exist at the time of the product's launch."

How are serious, potentially deadly adverse events from drugs going to be considered if the focus is "a keen eye toward the needs of the marketplace"? These companies aren't looking to fill a health need. They are anxious to design drugs that appear to fulfill a human desire. The aim of drug companies is to sell drugs as if they are candy instead of medicines to use when a person's health absolutely requires it.

"Today's research and development is influenced as much by commercial unmet needs as it is by clinical opportunities. . .Armed with this information your company can formulate winning lifestyle drug development." This is from a synopsis describing a Reuters Business Insight book titled *The Lifestyle Drugs Outlook to 2008: Unlocking New Value in Well-Being*. It is about designing diseases and marketing drugs for healthy people.

Do you believe marketing and advertising companies are considering your health "needs"? Advertising agencies are successful by playing to human desires and fears. People desire to be happy, young, sexy and admired. People fear diseases, growing old, and dying. Human *needs* are actually very simple. Basic needs are simple and have nothing to do with egotistic desires. Living simply and helping others doesn't make riches for advertising firms or pharmaceutical companies. They gain by playing to the human ego.

Wake up and smell the baloney! Eternally youthful life is not found in a drug. Of course for many people prescription drugs have brought eternal rest. In fact, more than 106,000 Americans die from prescription drugs every year. Almost twice the number of Americans who died in all the years of the Vietnam War are killed every year from prescription drugs.

False Marketing Leads to High Costs – In Lives and Money

Every day a tsunami of drug advertising crashes over millions of Americans. It is seen on television, in print and on the Internet. A high percentage of the advertising is false or misleading as demonstrated by the fact that between the years 2001 and 2005 the FDA sent 170 warning letters to 85 drug companies for misleading and/or inaccurate advertising for 150 drugs. Physicians were targeted by 62 percent of the false or misleading messages. The other 38 percent were misleading and inaccurate advertising on television and in print. Since the warning letters were not publicized, the public and physicians who heard or read the false or misleading information remained naively unaware and unprotected. The examples of dishonest advertising by Merck and Pfizer for their COX-2 inhibitor drugs is the tip of an immense and very frightening iceberg.

"If we put horse manure in a capsule, we could sell it to 95 percent of these doctors." Former CEO of Parke-Davis Harry Loynd's statement is not outlandish when you consider the hundreds of times drug marketing has won over true science. Expensive marketing campaigns for drugs continue to drown the truth about many prescription drugs. Loynd's slogan within his company was, "Pills are to sell, not to take."

You must protect yourself by not paying attention to drug ads. Don't be lured by the idea of a wonder pill. Why trust drug companies? Have their actions demonstrated they are trustworthy? Don't over-estimate what a chemical compound can bring you. For the vast majority of people, better health is achievable with a healthy lifestyle. The human form has existed more than 50,000 years because of its incredible immune system and ability to heal.

Ask your doctor about alternatives to drugs. Ask about risks of any drug your doctor recommends. If your doctor does not have time to talk with you, ask a pharmacist, call the FDA, or do some Internet research using websites that provide patient comments like http://www.askapatient.com. Keep in mind the FDA has cited drug firms for false marketing messages to physicians for misrepresenting risks of drugs, promoting unproven uses of drugs and making unsupported or misleading claims. In the FDA warning letters noted above, there were 82 times that marketing materials shown to doctors made false claims about the results of clinical trials. In some instances, the authentic clinical trial results actually contradicted the marketing claims. Figuratively speaking, the drug sales people were trying to sell physicians capsules of horse manure.

Drug salespeople are paid to get physicians to write more prescriptions. The immense army of 100,000 people are paid to go out every day to convince physicians to prescribe new, expensive drugs when older, less expensive drugs may be just as effective or better yet, life-style changes or non-drug therapies may be the best remedy for better health. Non-drug therapies are not well understood by most physicians in our current drug-centered medical environment. People who died from taking Vioxx may still be alive today if they had used alternative methods for their pain relief.

Herbalists help Swiss, French, Germans and other Europeans and Asians use natural herbs to promote healing. In a National Institute of Health (NIH) study completed in 2002 about major depression, the people in the study with the best results, 32 percent, were those taking the placebo. A placebo has no active ingredients. It is given to a control group during clinical trials to get the results of a non-medicated group, demonstrating that the body-mind has its own capacity for healing. The herb St. John's Wart provided full response to treatment for depression in 24 percent of patients without any side effects.

Simple life changes have been found to be extremely effective to improve your sense of well-being, such as increased sleep, regular exercise, less television watching, healthy eating, owning a pet, and laughing. Yes, laughing. When we laugh, the body releases natural chemicals called endorphins that relieve pain and improve our sense of well-being. Laughter is truly the best medicine.

Get to the Numbers That Count

If you were told a drug reduced the risk of a disease by 50 percent would you be impressed? It does sound like a significant reduction. Statistics are numbers that need to be taken in context. Drug companies and their advertising agencies know how to use statistics to catch your attention and the attention of physicians. You do not have to be a statistician to uncover the true numbers. Let me show you how to find the most informative numbers so you will not be bamboozled by drug ads with striking percentages.

Who was the group in the clinical trials? Merck advertised their drug Fosamax (for bone density) as reducing the risk of hip fractures by 50 percent. Merck ads never mention that this number comes from clinical trials on older women who had already experienced one or more fractures. In other words, for the millions of healthy women the ads are aimed at, there is no way of knowing if there is any actual benefit from taking the drug.

Looking further at the real numbers that count we find that in this elderly group of women in the study, 2 out of 100 of those taking the placebo experienced a hip fracture during the study. So 2 percent of the control group broke their hip. In the group taking Fosamax 1 out of 100 experienced a hip fracture during the study. One percent of the group taking the drug broke their hip. That is only a **one percent difference in**

absolute numbers. Since one is half of two, the 50 percent change is not false, it is expressing in relative terms, not the absolute terms of one percent. The absolute is not impressive, which is why the advertisements did not give the real numbers that count. Merck advertises Fosamax to all women using fear of weak and breaking bones to sell their drug.

No benefit for healthy younger women was the result of a much larger study of Fosamax conducted by the government. But that does not stop drug salespeople from encouraging physicians to prescribe the drug for all sorts of women. And you can bet they use the 50 percent figure to sell it to doctors.

Canadian researchers and physicians from the British Columbia Office of Health Technology Assessment reviewed all the scientific data about osteoporosis and presented their findings at a meeting of the American Sociological Association. The title was, "Normal Bone Mass, Aging Bodies, Marketing of Fear: Bone Mineral Density Screening of Well Women." These independent experts found the extensive use of bone density testing was another case of marketing to women's fear.

Admired movie star Sallie Field was at the 2007 U.S. Women's Golf Open to promote osteoporosis as a disease and to advertise Boniva, a drug developed by Roche Laboratories. Roche and GlaxoSmithKline (GSK) partner to market Boniva. Paying Sallie Field to advertise the drug draws much public attention.

Dr. David Henry, head of a multi-disciplinary team of public health scientists and medical doctors at the University of Newcastle, Australia, is concerned osteoporosis is being defined by the drug industry as a disease that needs to be treated with drugs. "That is disease mongering," says Dr. Henry.

Drug companies are creating the illusion that millions of women suffer or will suffer from osteoporosis. Merck has even reported to stockholders that less than a quarter of the market has been diagnosed and treated. When you hear advertising from any source expressing that millions of people are undiagnosed and untreated for a disease, realize you are not being given the true numbers. You are simply hearing a drug company creating fear to attempt to enlarge the market for a drug.

"Research evidence does not support either whole population or selective bone mineral density testing of well women at or near menopause as a means to predict future fractures" was the conclusion of the Canadian analysis of the entire body of evidence about testing for women's bone density.

"Excruciating, debilitating hip, back and leg pain." Wrote a woman about her side effects from Boniva. "This stuff is lethal. Do not take it unless you want to die!" another woman warns. Hundreds of women have written comments about their serious side effects from taking Boniva on the website http://www.askapatient.com. The website is helpful because some women have reported the side effects to their physicians and have been told it couldn't be from the drug. Some women were put on additional drugs to treat the side effects (adverse events) they suffered from Boniva. Other women simply stopped taking Boniva and the symptoms stopped, as they reported on the website.

When you are quite certain a drug is causing serious side effects tell your doctor you are going to stop taking it. Going on a multi-drug routine to treat the side effects of a drug is not getting to the root cause of the health issue. If your physician doesn't believe you are suffering side effects, do your own research on the Internet. You can check most drugs on askapatient.com. Remember the drug companies are often not telling physicians about severe side effects. Trust what your body tells you. And report your adverse event to the FDA directly. Your health is your personal responsibility.

Some women taking Merck's Fosamax have suffered the adverse effect of Osteonecrosis of the jaw, also called Dead Jaw. The jawbone decays and dies, causing exposed portions of jawbone inside the mouth. Osteonecrosis often requires surgery and can be disfiguring.

Don't be fooled by relative term percentages. If a drug is said to reduce your risk of a stroke by 33 percent it is the same as the drug reducing the risk of stroke from 3 percent down to 2 percent. That is, 3 percent of the placebo group had strokes and 2 percent of the group taking a drug had strokes. Because 3 minus 2 is 1, in absolute terms it is only one percent reduction of risk. The reduced risk by 1/3 or 33 percent is in relative terms. It can also be said that out of every 100 people taking the drug over the time of the study, one person will be saved from having a stroke. That is the one percent.

This is also known as the **number needed to treat (NNT).** One hundred people (with a certain health profile) are needed to take the drug (over a recorded period of time) to benefit one person. If two people out of one hundred have reduced risk, the number needed to treat is fifty. Out of every 50 people (with a certain health profile) one person has reduced risk. There is no way of knowing which person.

Would you buy and take a drug for years or the rest of your life for those results? It is also important before you make the decision to take a drug, to find out if there are life-style changes that may provide the health benefits without the drug. It is completely your choice. These are the numbers you want to uncover before you assume you will benefit from taking a drug. The number needed to treat is important to ask for. Also ask about the health profile of the people used in the clinical studies. If the people taking the drugs in clinical studies had specific health conditions you do not have, there is no reason to assume you will benefit from taking the drug.

Healthy Life Expectancy

If you were asked where the U.S. ranked in the world for healthy life expectancy, what would you respond? First? Second? Third? You probably would not guess 28th. That is however where the U.S. ranked in the World Health Organization's Healthy Life Expectancy for the year 2000. The U.S. population is less than 5 percent of the world's total, yet it consumes almost 50 percent of all prescription drugs. We are the most medicated nation, but far from the healthiest. Other countries with better healthy life expectancy than the U.S. are living without the deluge of billions of dollars a year of drug advertising.

Drug dependency has a high price. It is costly in lives and income. The public does not benefit from the tsunami of drug advertising washing over them every day. Physicians do not need legions of drug representatives interrupting their work and misrepresenting the truth. It's time to just say no to drug companies.

10

CANCER IS BIG, BIG BUSINESS

When a person is diagnosed with cancer, s/he is suddenly worth $200,000 to $600,000 to the U.S. medical system. Much of that goes to the pharmaceutical companies for chemotherapy drugs; then for the drugs to overcome the adverse events of chemotherapy; until the patient dies from chemotherapy.

> "Chemotherapy does not eliminate breast, colon, or lung cancers. This fact has been documented for over a decade, yet doctors still use chemotherapy for these tumors."
>
> Allen Levin, M.D. *The Healing of Cancer.*

> "…as a chemist trained to interpret data, it is incomprehensible to me that physicians can ignore the clear evidence that chemotherapy does much, much more harm than good."
>
> Alan C. Nixon, Ph.D., former president of the American Chemical Society.
> "Patients are as well, or better off untreated." Stated Hardin Jones, M.D. Professor of Medical Physics and Physiology at the University of California, Berkeley. Dr. Jones analyzed cancer survival statistics for decades.
> Dr. Ulrich Abel, a German epidemiologist and biostatistician, reviewed cancer articles from all the leading medical centers and journals and published his findings in the Lancet, August 1991. "Success of most chemotherapies is appalling. There is no scientific evidence for its ability to extend in any appreciable

way the lives of patients suffering from the most common organic cancers...Chemotherapy for malignancies too advanced for surgery, which accounts for 80 percent of all cancers, is a scientific wasteland."

Cancer treatment is big, BIG money; nearly $100 billion spent in 2006 according to a study published in the *Journal of Clinical Oncology*. Chemotherapy is a sizable chunk of that amount and the drugs are becoming more expensive. Avastin, approved in 2004, lists for about $4,400 per month. Tarceva is about $2,000 a month. Sometimes oncologists want to use both drugs on a patient. How can Medicare or Medicaid be expected to pay more than $6000 a month just for drugs for one person? When the lack of success of chemotherapies for most cancers is factored in we are talking incomprehensible waste.

Tykerb, FDA-approved in March 2007, is another chemo drug with a hefty price of $2,900 per month. It is to be used in combination with an older chemotherapy, Xeloda, for advanced breast cancer patients. What may the patient gain? The study results were an additional two months of delayed tumor growth and living with the side effects of two highly toxic drugs. Mortality rates did not change.

It is utter madness. Insanity has been described as doing the same thing over and over and expecting different results. For fifty years drug companies have focused on reducing the size of tumors with highly toxic drugs. Tumor size is easily measurable. It is the type of research that the FDA promotes. But tumor regression is an ineffective predictor of a cancer treatment's ability to thwart progression of the disease in the human body. Tumor regression is not providing realistic results for the vast majority of cancer victims. And the price tag continues to climb.

The Great Lie

"In the end, there is no proof that chemotherapy actually extends life in the vast majority of cases, and this is the great lie about chemotherapy, that somehow there is a correlation between shrinking a tumor and extending the life of a patient." Ralph Moss, Ph.D. said in a British television interview in 1994. Dr. Moss has independently evaluated the claims of non-conventional and conventional cancer treatments for more than 30 years. He has authored twelve books about cancer and alternative treatments, including *Antioxidants Against Cancer, Questioning Chemotherapy* and *The Cancer Industry.* In July 2007 Dr. Moss was appointed to the Scientific Advisory Board of Breast Cancer Action.

Early in his career he was fired from his job at the Sloan Kettering Cancer Institute because he spoke out publicly about the Institute's cover-up of the positive results of cancer treatment with amygdaline. Dr. Kanematsu Sugiura, a senior researcher, had shown Dr. Moss the research results confirming amygdaline was extremely effective in stopping metastasis, the spread of cancer. It is cancer's ability to spread beyond the immediate area that makes it so deadly. Amygdaline is a natural product in fruits, especially apricot seeds inside the pit. Crack an apricot pit and inside is a seed that looks like an almond. It can be eaten raw and tastes like a bitter almond. Amygdalin is sometimes called laetrile. It has also been called vitamin B17 however Amygdalin is not a vitamin. There is a U.S. patented Laetrile, a partly synthetic substance. Mexico produces a different laetrile that originates from ground apricot seeds.

There are other natural plants shown to be natural cancer fighting agents. The plant known as Cat's Claw was reported in 2001 by Italian researchers to directly inhibit the growth of breast cancer cells by an amazing 90 percent. Swedish researchers found Cat's Claw inhibited lymphoma and leukemia cell growth. Cat's Claw originates from the bark of a rainforest tree.

Harmful Cancer Screening

Mammograms harm 10 women for every 1 woman helped was the recent findings by the Nordic Cochrane Center in Denmark. The Center sets the gold standard for cancer research in Scandinavian countries. The Center reviewed all trials of mammography screening. More than 500,000 women's outcomes were reviewed. The absolute risk reduction for breast cancer death in women that had regular mammograms was 0.05 percent. This means for every 2000 women getting mammograms over a ten-year period, one woman would have her life prolonged. And 10 women would be misdiagnosed and receive unnecessary chemotherapy, radiation and surgery. Denmark has discontinued using mammograms for breast cancer screening.

Thomas Edison completed work on the first X-ray machine in 1904. His assistant, Clarence Dally, began work with Edison on the development of Edison's X-ray focus tube in 1895. By 1900 Dally had cancer in both hands and in 1902 both of his arms were amputated but he died of cancer in 1904. Edison said, "The X-ray had affected poisonously my assistant, Mr. Dally." Edison abandoned his research of X-rays upon Dally's death and refused to ever have an X-ray on himself.

Women have been led to believe the low energy X-rays of mammograms are safe, but radiation research findings in 2002 from the National Institute of Health found that low energy X-rays of mammograms produce an increased biological risk, as opposed to higher energy photons of other types of X-rays. Exposure to ionizing radiation, the type used in mammograms, creates a wound that promotes a microenvironment in the tissue surrounding breast cells that can cause cells to become cancerous. Cell biologist Mary Helen Barcellos-Hoff of the Lawrence Berkeley National Laboratory explains, "Our studies demonstrate that radiation elicits rapid and persistent global alterations in the mammary gland microenvironment. Radiation exposure can cause breast cancer by pathways other than genetic mutations (DNA damage).

Ionizing radiation is a well-established carcinogen because it damages DNA. DNA is the body's building plan for all cells. German research revealed that exposure to low-dose radiation causes damage so extensive to the DNA that it is unable to repair itself, resulting in permanent genetic mutation of the DNA.

The push to use mammograms for breast cancer detection in the U.S. began in the 1970's. Since then, the percentage of women diagnosed with breast cancer has tripled. Many of the women are misdiagnosed and many others may have developed cancer because of the radiation from mammograms.

Vitamin D3 was found to reduce the risk of breast cancer and all other cancers by an astounding 77 percent. It appears that money would be much better spent on cancer prevention education rather than the emphasis on questionable cancer screening. Mammogram screening is part of the huge income generating cancer industry. Cancer prevention is a threat to the many factions of the cancer industry, including pharmaceutical companies. Political appointees at government agencies like the National Institute of Health opposed their own expert panel findings that mammograms are not beneficial for women less than 50 years old.

People who gain from the cancer industry feel they have something to lose if people take action to prevent cancer. Women, on the other hand, have nothing to gain from mammograms and everything to lose.

Research of Metastisis is Needed

Metastasis causes death in 90 percent of cancer patients. And yet less than one-half of one percent of all cancer research proposals since 1972 have focused on metastasis. National Cancer Institute in 2006 funded nearly 8,900 studies. Metastasis was not even mentioned in 92 percent of all of these research proposals. The established cancer research system in the U.S. is stuck in a pattern that does not allow innovative thinking. That is exactly the conclusion of a panel of cancer center directors meeting in February 2003. They concluded that without major change, "the system is likely to remain inefficient, unresponsive, and unduly expensive."

The system continues to depend on the reductionist model, dividing the whole into smaller and smaller parts, focusing attention on parts as though they exist independent of the whole. The human form is an integrated system of spirit-body-mind functioning with an intricate energy-communication flow. The reductionist model of our current drug-centered medical system ignores the absolute fact that any change affects the whole. The human form is designed to bring itself back into balance. It is a living integrated matrix. The war on cancer is a violent, destructive attitude. When we wage war on a tumor with toxic poisons we are also waging war on the entire human form, damaging the form's ability to ingest nutrients, debilitating the miraculous immune system, and damaging all tissues.

"We do not need to fear biological warfare. We are doing it to ourselves." James Oschman, Ph.D., author of *"Energy Medicine, The Scientific Basis."* Dr. Oschman's book provides the scientific evidence of why complementary and alternative medicine (CAM) is grounded in science. Energy medicine research deserves the attention of the mainstream medical culture. Unless innovative approaches to cancer treatments are introduced, the death rates of cancers will not diminish.

That is precisely the conclusion of a major study published in the *British Medical Journal* in August 2006. The clinical study results of 12 new anticancer drugs approved for the European market from 1995 to 2000 were compared to the older, standard drugs for their respective cancers. The only thing that went up was the expense of the drugs. There was no improved survival rate with the new drugs. There was no better quality of life. There was no better safety. No advantages at all, just much higher price tags on all 12 of the new drugs. In the case of one drug the price was 350 times higher than the older, standard drug. We can only hope all oncologists (physicians specializing in cancer treatments) read the *British Medical Journal* article before the drug salespeople arrive.

Why does the FDA continue to approve drugs that do not offer improved survival rates and are hundreds of times more expensive than the old drugs? As Dr. Oschman said at the University of North Carolina Complementary and Alternative Medical (CAM) Conference in April 2005, "When a technique produces cash flow it becomes institutionalized." Chemotherapy certainly produces cash flow. Taxol is a chemotherapy drug that made $9 billion from 1994 through 2002.

Erbitux was approved by the FDA in February 2006 for cancer of the colon and rectum. It did shrink tumors in the clinical trials but patients' survival rate was not improved. A weekly dose costs $2,400. So patients will still die but their families will be left with even greater financial burdens. Is that progress?

The opportunity to make huge monetary profits from chemotherapy has put some drug firms on the wrong side of the law. Like the EPO drugs for severe anemia mentioned in chapter four, drug companies sometimes encourage physicians to make additional income from selling chemotherapy drugs. TAP Pharmaceuticals was required to pay $885 million to the federal government to settle criminal and civil charges for drug fraud in 2001. And another $150 million to settle a class action suit for the cancer patients' out of pocket expense. The company was accused of bribing physicians with discounts to prescribe their cancer drug, Lupron, and suggesting the doctors could then charge Medicare and Medicaid the full reimbursement price and pocket the difference. The company's sales force carried out the plan. Do we really need door-to-door selling of drugs to physicians?

Cost of Adverse Events from Chemotherapy

Chemotherapy drugs are so toxic that the *Handbook of Cancer Chemotherapy* for medical personnel lists sixteen OSHA safety procedures. The Environmental Protection Agency regulates disposal of needles and other equipment used with these dangerous drugs. The Handbook warns medical personnel who handle chemotherapy drugs:

> "The potential risks involved in handling cytotoxic agents have become a concern for healthcare workers. The literature report various symptoms such as eye, membrane, and skin irritation, as well as dizziness, nausea, and headaches experienced by healthcare workers not using safe handling precautions. In addition, increased concerns regarding the mutagenesis and teratogenesis (deformed babies) continue to be investigated. Many chemotherapy agents, the alkylating agents in particular, are known to be carcinogenic (cancer-causing) in therapeutic doses."

These are the drugs that are injected directly into the blood of patients sick with cancer. The bloodstream carries the toxic drugs to all areas of the body. Most chemotherapy does not distinguish between cancer cells and normal cells. That is why since President Nixon declared the "war on cancer" in 1971 we have not seen any appreciable difference in the survival rate of the majority of cancers.

"Chemotherapy-related serious adverse effects among younger, commercially insured women with breast cancer may be more common than reported by large clinical trials and lead to more patient suffering and healthcare expenditures than previously estimated." A review of more than 12,000 women younger than 64 years old who were newly diagnosed with breast cancer concluded. Within the 12 months after initial diagnosis of breast cancer, 61 percent of the women who received chemotherapy were hospitalized or visited the emergency room. This added an average of more than $18,000 per woman annually to the medical expenses. Fortunately for these women, they were insuranced.

In the case of women who received chemotherapy for ovarian cancer, direct medical costs for adverse events due to chemotherapy ranged from $688 to $7,546 per episode of care according to another study assessing the cost of medical care for chemotherapy adverse events.

Besides the additional financial expenses of chemotherapy adverse events, there are the additional health problems for people whose systems are already compromised by cancer. A recent study found that chemotherapy changes the way the brain works in some patients. PET scans showed the functioning of the frontal cortex had changed in breast cancer patients who had received chemotherapy. This affected short-term memory. Researchers have named this adverse event "chemobrain". Chemobrain is permanent.

Irreversible hearing loss from chemotherapy is a condition caused from any of several platinum-based chemotherapy drugs used to treat brain tumors in children. A study published in the *Journal of Clinic Oncology* in 2006 found that the severity and frequency of hearing loss due to these drugs (called ototoxicity) is under-reported because it is difficult to diagnose unless a physician is aware and tests for it. The pediatric audiologist (children hearing specialist) who led the study explained that ototoxicity often is not diagnosed because it is not total hearing loss. It is progressive loss of hearing of higher frequencies. Also, young children will not complain when they don't understand what they have heard. The problem is they have not heard all the frequencies so they are missing sounds that are part of the message. The inability to hear everything affects educational performance and social-emotional development. A 1998 study of 1200 children with minimal hearing loss found that 37 percent failed one or more grades in school compared with the rate of 3 percent for children with no hearing loss.

Cancer is a side effect of chemotherapy as the *Handbook of Cancer Chemotherapy* warns medical personnel. Who warns the patients? Chemotherapy may cause some cancers to metastasis to other areas of the body, creating a situation where the patient endures the chemo treatments, is told all looks good and then a few months later cancer is found in another area and another round of chemo is recommended. The immune system is the body's natural protective means to rid the body of foreign cells. Chemotherapy suppresses the immune system. Children who receive chemotherapy for Hodgkin's disease are 18 times more likely to later develop secondary malignant tumors.

In a survey 75 percent of the responding **oncologists said if they had cancer they would not participate in chemotherapy trials because of chemotherapies' "ineffectiveness and its unacceptable toxicity."**

Even so, oncologists treat 75 percent of all cancer patients with chemotherapy. Why? They do not know of any other choices in our drug-centered medical system. These physicians want to do something for people with cancer. They make their living by treating cancer patients. Treating cancer in our medical system has three options: chemotherapy, radiation, and surgery. Thinking outside the drug-centered box is not acceptable in the current system. Patients and their families are desperate for hope. It seems like something as destructive as chemotherapy should "destroy" the cancer. Unfortunately it is not that simple.

If You Want Healthcare Choice, Consider Moving

America, land of the free? Not when families are forced to have their children receive chemotherapy and radiation against their wishes. In the U.S. today, parents do not have the right to choose what types of cancer treatment their children receive.

In May 2009, thirteen-year old Danny Hauser was ordered by the court to submit to chemotherapy and radiation treatment for Hodgkin's lymphoma. Danny had received chemo earlier in the year and became extremely ill. He wanted to use natural methods to cure his cancer rather than take any more traditional cancer treatments. An international police search was initiated when Danny and his mother tried to escape to Mexico. When the fugitives returned to their home in Minnesota, a district court judge ordered that Danny's parents lose custody of their son. The judge allowed the parents to retain custody only after they agreed to have their son receive chemotherapy. Danny's website, dannyhauser.com, provides insight to Danny's anger and emotional trauma.

The physicians treating Danny claim he would die without the toxic treatments. Billy Best had heard the same prediction when he was only sixteen-years old. Diagnosed with

Hodgkin's lymphoma, Billy wanted to use natural cures. His mother cried when she read the list of side effects that chemo could cause. She was required to sign a statement that she would not hold the doctors responsible if Billy got cancer from the chemotherapy. Billy ran away from home and hid from his family and the law in order to avoid chemotherapy. He lived on a diet of raw vegetables, fruits, roots and medicinal herbs, avoiding meat, dairy, and sugar. He also used the natural cancer therapy 714-X from Canada. The natural treatment cured him. As of May 2009, he is thirty-one years old, healthy, and cancer-free; making the physicians' predictions of his imminent death a ridiculous fallacy.

There have been other cases of forced chemo and radiation on children. In 2006, Virginia social services took the family of Abraham Cherrix, age 16, to court where Abraham's parents were ordered to treat his cancer with chemotherapy. When Abraham was first diagnosed with cancer he did receive chemotherapy and it left him so weak and ill he chose not to receive any additional chemotherapy. He wanted to treat his cancer with alternative methods that would strengthen his immune system. Besides having the trauma of a teenage son with cancer, Abraham's parents had to hire attorneys and fight the state twice because the first judge found the parents medically negligent and ordered Abraham to take more chemotherapy. The appeal judge overturned the finding but did order radiation therapy along with the alternative methods that Abraham had chosen. Abraham could not just focus on overcoming cancer. He had the stress of fighting a legal battle. In both ways he was fighting for his life.

The state of Texas forced Katie Wernecke to receive chemotherapy and radiation. Katie and her parents had wanted to treat Katie's Hodgkin's lymphoma with immune-supportive alternative choices. Her mother was jailed when she took 13-

year-old Katie into hiding in Texas to avoid the toxic treatments. Katie was then given the chemo and radiation under duress.

According to the Mayo Clinic website Hodgkin's lymphoma treated by radiation increases the risk of other forms of cancer, especially for girls. "The risk of breast cancer from standard dose radiation is even higher for girls and women treated when they are younger than 30 years. The risk is generally too high for this therapy to be considered." The Mayo Clinic website also explains a major concern with chemotherapy in the treatment of Hodgkin's lymphoma is the possibility of long-term side effects such as heart damage, lung damage, liver damage and secondary cancers, such as leukemia.

The pharmaceutical and medical establishments will try to crucify anyone who speaks out about recovering from cancer using natural cures, as is the case of former orthopedic surgeon Dr. Lorraine Day. Dr. Day had breast cancer two decades ago. She refused traditional treatments because of their extreme side effects and inability to cure. She healed herself through natural means and now helps others to understand how to cure cancer and other diseases using simple, natural means. Dr. Day continues to be criticized by those who profit from traditional western medical treatments.

Vaccine Hoax That Uses Cancer Scare Tactics

As drug companies continue to contribute large sums to politicians, we will see more legal battles for health choice rights. The FDA approved in June 2006 a vaccine reported by Merck to prevent cervical pre-cancers and non-invasive cervical cancers associated with four strains of HPV. The study explains the vaccine Gardasil, "does not prevent all HPV types associated with cervical cancer." There are more than 100 strains of HPV.

According to an FDA document, HPV infections do not lead to cervical cancer. "Most infections by HPV are short-

lived and not associated with cervical cancer," states an FDA news release dated March 31, 2003. And according to another FDA document, **Gardasil has been found to increase the risk of developing precancerous lesions** by 44.6 percent in women already infected with the HPV strains used in the vaccine.

What are the facts about cervical cancer?

1) It can occur in women who do not have any type of HPV.
2) It is a cancer that can be diagnosed early with annual PAP smears.
3) It is easily treatable when found early, without chemo, radiation or surgery.
4) It is not contagious.
5) In the U.S. the annual death rate from cervical cancer is 2 to 4 women per 100,000 women.

The pharmaceutical industry has strong financial ties to some politicians. Texas governor, Rick Perry, in February 2007 bypassed the state legislature and announced an executive order requiring girls age 12 and older to receive Gardasil vaccinations. In May 2007 the Texas legislature responded to public outcry and created a bill that would not allow mandated vaccination.

Cost of the HPV vaccine in 2007 was approximately $360. If all girls ages 12 to 18 years old in the state of Texas were required to get the vaccine Merck would make a substantial profit from just one state. And the girls would be put at risk of death or severe side effects including developing cervical cancer.

Thirty-two deaths and more than 12,400 serious adverse events have been reported to the FDA in the first three years the vaccine has been on the market. Some of the side effects young women have suffered include dizziness, nausea, headaches,

Guillan-Barre syndrome - a disorder that causes the body's immune system to attack the nervous system sometimes causing paralysis, anaphylaxis - which can cause sudden death, blood clots, and pancreatitis.

Ashley Ryburn's life as a healthy, active teenager changed completely when she received the Gardasil vaccine. She is now on several drugs and suffers from extreme pain. She had to quit all sports activities and cannot even ride her bicycle around the small town where her family lives. Ashley's video story can be seen at http://www.rockymountainnews.com /videos/detail/ashley-story/.

Physician Scott Ratner's daughter became seriously ill with a chronic autoimmune disease after her first dose of Gardasil. Dr. Ratner told CBS News, "she'd have been better off getting cervical cancer than the vaccine."

Norwegian physician Charlotte Haug, M.D., Ph.D., M.Sc., stated her concerns about the Gardasil vaccine in an editorial in the *Journal of American Medical Association*. "Whether a risk is worth taking depends not only on the absolute risk, but on the relationship between potential risk and the potential benefit. If the potential benefits are substantial most individuals would be willing to accept the risks. But the net benefit of the HPV vaccine to a woman is uncertain. Even if persistently infected with HPV, a woman most likely will not develop cancer if she is regularly screened."

Merck Continues To Seduce States With Money

Time Magazine reports that Merck has set aside an undisclosed amount of money specifically to lobby states for mandatory HPV vaccinations. It seems to be working. As of May 2007, the legislatures of 18 states were considering mandatory Gardasil legislation.

"It is silly to mandate vaccination of 11 to 12- year-old girls," states Dr. Diane M. Harper, director of the Gynecologic Cancer Prevention Research Group at Dartmouth Medical School. "Giving it to 11 year-olds is a great big public health experiment." Dr. Harper is internationally recognized for her 20 years of research of the more than 100 strains of HPV.

Wake up America and smell the baloney. These are your children. If you think the government is looking out for your children's good, you have something to learn from the next chapter. If the government and pharmaceutical companies are putting the good health of children as a priority, why does the **U.S. infant mortality rate rank 42nd in the world**? That is correct, 42nd in the world according to the CIA World Fact Book. We have the most expensive and drug-centered medical system in the world. And 41 other countries have healthier babies, including Cuba.

It is not surprising given the lack of federal government concern for children. For example, the Environmental Protection Agency (EPA) allows **pesticide testing on "orphaned, neglected, abused and mentally handicapped children." And no adult consent is required.** That was the EPA ruling as of September 2007. The EPA received more than 50,000 letters of protest from members of Congress, medical associations, the public, and protests from EPA scientists. But the political leaders of the EPA, with their financial ties to industry, ignored all and made the most vulnerable U.S. citizens, those who need the greatest protection - the children without loving parents to speak for them, available as test subjects for chemical companies.

Would Abraham Lincoln recognize this country? Would he still speak of it as a government of the people, by the people, for the people?

11

PEOPLE WITHOUT CONSCIENCE

Mad as a hatter. The phrase originated in the English Victorian era because hat makers often went insane due to the handling of mercury in the process of making hats. It has been recognized for more than one hundred years that mercury causes brain damage.

So why do U.S. pharmaceutical firms use a mercury-based preservative for vaccines given to babies and young children? It saves money. Thimerosal developed by Eli Lilly in the 1930's is a preservative agent to prevent bacterial and fungal contamination. It also allows multiple injections to be made from one vial of vaccine. One large vial to give many shots is less expensive than bottling vaccines in small individual doses. The reality is Thimerosal use in vaccines is unimaginably more expensive than the non-mercury alternatives. Not to the drug companies. But for the millions of families who now have children with autism or neurological disorders due to mercury poisoning. Maybe the updated American term should be: Mad as a drug company.

Unfortunately for American families the madness was even more widespread than the drug companies. In 1997 Congress passed a resolution requiring the FDA to review mercury in drugs and biologics. The meeting to review findings resulting from the Congressional mandate was held secretly in June 2000. Under federal "Sunshine Laws" a public announcement should have been made in advance of the meeting. No such announcement was made and the meeting notes were withheld from the public until 2003. Access to the database was withheld until 2005. As of September 2007 only one independent scientist has been allowed to study the data. The embargo of the meeting notes and data appear to be in direct violation of the federal Freedom of Information Act.

The secret meeting held at Simpsonwood was organized by the Centers for Disease Control and Prevention (CDC). The CDC went to great lengths to ensure everything discussed in the meeting remained totally secret. Meeting participants included high level officials from the FDA and CDC, vaccine specialists from the World Health Organization, and representatives from each of the major vaccine manufacturers: Aventis Pasteur, Glaxo-SmithKline, Merck, and Wyeth. Throughout the meeting the CDC reminded everyone no copies of any documents could be made and no papers could be taken from the meeting. No paper trails and sadly no whistleblowers.

The spy-like secretiveness of the meeting was due to the frightening news that a mercury-based preservative, Thimerosal, used in children's vaccines was responsible for autism in the U.S. to increase to the staggering ratio of one in every 150 children. The CDC epidemiologist who had analyzed the database of medical records for 100,000 children admitted he was stunned. According to the meeting minutes eventually released, Dr. Tom Verstraeten told the assembled group, "When I saw this, and I went back through the literature, I was actually stunned by what I saw because I thought it plausible." (*relationship of Thimerosal and autism which his analysis uncovered*) He goes on to cite the many earlier studies done by others that had found a positive relationship between Thimerosal and autism, attention-deficit disorder, speech delays, and hyperactivity. All of these are neurological development disorders.

The American Academy of Pediatrics was represented by Dr. Bill Weil. Dr. Weil told the group, "The number of dose related relationships are linear and statistically significant. You can play with this all you want. They are linear. They are statistically significant...I work in the school system where my effort is entirely in special education and I have to say that the number of kids getting help in special education is growing nationally and state by state at a rate we have not seen before."

Instead of releasing the findings immediately to protect American babies and families, this group of government officials and industry experts discussed how to keep the news from the public. "If an allegation was made that a child's neurobehavioral findings were caused by Thimerosal containing vaccines, you could readily find a junk scientist who would support the claim with a reasonable degree of certainty. But you will not find a scientist with any integrity who would say the reverse with the data that is available. And that is true. So we are in a bad position from the standpoint of defending any lawsuits if they were initiated, and I am concerned." Said Dr. Robert Brent, a pediatrician at the Alfred I. duPont Hospital for Children in Deleware. According to what Dr. Brent said, a scientist with integrity would see the relationship between autism and Thimerosal just as a "junk scientist" would. Dr. Brent's concern certainly wasn't about preventing any more children from the harm of Thimerosal.

And Dr. John Clements, vaccine advisor at the World Health Organization, gave it all away when he announced, "And I really want to risk offending everyone in the room by saying that perhaps this study should not have been done at all because **the outcome of it could have, to some extent, been predicted.**"

Predictable? Absolutely. In 1999 Dr. Verstraeten had apprised FDA and CDC officials that infants were receiving up to 100 times the mercury exposure considered safe for adults by the Environmental Protection Agency (EPA). Mad as a drug company? Madness denotes mental deficiency and these men knew exactly what they were doing. They intentionally allowed hundreds of thousands more children in the U.S. to be permanently neurologically impaired by Thimerosal after their secret meeting as they sought to ensure they were not held responsible.

Dr. Verstraeten's study which was to have been published in a scientific journal, was not allowed to be published. Dr. Roger Bernier, associate director for science for the CDC's national immunization program, announced at the meeting, "We have asked you to **keep this information confidential**...If we could consider these data in a certain protected environment. So we are asking people who have done a great job protecting this information up until now to continue to do so." And so they did. No leaks. No whistleblowers. The vaccination program that had exponentially increased the number of vaccines with mercury given to infants based on CDC guidelines continued to be applied without warnings to physicians or parents. Babies and young children with forming brains and neurological systems continued to be exposed to as much as 100 times the amount of mercury the EPA defined as safe for adults. Those three years of secrecy caused hundreds of thousands more children to be permanently brain and neurologically damaged because these men wouldn't take responsibility for their actions. They chose to protect themselves, not the babies.

After the secret meeting the government database, developed with taxpayers' money, was turned over to a private company, America's Health Insurance Plans. The CDC had the data stored as off-limits to any researchers. Putting the database with a private company prevented researchers from gaining access to the database through the Freedom of Information Act, which makes government databases public record. The government officials and corporate executives at the covert Simpsonwood meeting were determined the research findings would not go beyond the meeting even though it was a Congressional mandate that had initiated the research. These are people entrusted with the public's safety in the U.S. and worldwide.

Even more unbelievable is the FDA and CDC continue to allow this toxic, baby-maiming agent to be used in several types of vaccines here in the U.S. including flu vaccine, tetanus-diptheria (DTaP), and meningitis vaccine. Vaccines with Thimerosal continue to be shipped to poor third world nations where autism had been virtually unknown and is now becoming an epidemic. These are men without conscience.

Although children have been receiving vaccines containing Thimerosal for decades two significant changes created an entire generation of neurologically damaged babies. The CDC, like the FDA, has strong financial ties to the drug industry. The CDC's Advisory Committee on Immunization Practices (ACIP) sets vaccine policy. At the time of the Simpsonwood meeting, half the ACIP members were employees or consultants to the vaccine makers. This group changed vaccine policies in 1990, drastically increasing the number of vaccines and recommending the vaccines be given to infants. More vaccines to babies meant more money to drug companies. Besides the mercury, remember that every vaccine contains viruses. How sensible is it to infect infants with dozens of viruses?

Before 1990 preschool children (not infants) received only 3 vaccinations: polio vaccine which never contained mercury; DTaP vaccine is a mixture of 3 vaccines against diptheria, tetanus and pertussis; and MMR vaccine is a mixture of 3 vaccines for measles, mumps and rubella. DTaP and MMR contain Thimerosal. The baby-boomers as children received even fewer vaccines and at older ages.

Then in 1990 the number of immunizations given to babies and young children skyrocketed. The push for the increases came from the ACIP whose members were primarily from vaccine manufacturers or had close financial ties with them. These masses of vaccines were given at younger ages. Many children were receiving 22 immunizations before the age of 6 years. Most of these vaccines contained the mercury preservative. Some infants were being immunized at birth and 2 months, based on ACIP policy. One of the vaccinations recommended given within 24 hours of birth is for Hepatitis B, primarily a sexually transmitted disease. The vaccine contains Thimerosal.

Parents are not informed when there is mercury in vaccines so they can make a parental choice if they want to subject their babies to mercury to have the vaccinations. Unfortunately Americans have been lulled into thinking substances are safe if the government has given a stamp of approval. Babies with forming neurological systems and forming brains were subjected to mercury amounts far beyond what is found in tuna. Yet pregnant women are warned not to eat tuna because of the mercury in the fish due to environmental pollution.

Autism is a human created disorder. Thimerosal was first used in vaccines in 1931 and the first cases of autism were diagnosed in 1943. Besides autism many other types of neurological disorders and mental retardation are associated with Thimerosal.

The earliest research showed Thimerosal was dangerous and could cause death. In 1935 the vaccine manufacturer Pittman-Moore tested Thimerosal-based vaccines on dogs. Half the dogs died. Pittman-Moore researchers concluded

Thimerosal was not acceptable to use in dog vaccines. But that did not stop Eli Lilly. So an agent confirmed more than 70 years ago as too dangerous to be used in dog vaccines continues to be used in 2009 in vaccines for babies and children.

Pregnant women and their unborn children are not protected from mercury. Investigative reporter Evelyn Pringle recently discovered the injection given to pregnant women to stop contractions and prevent early labor contains Thimerosal. As does the injection given to pregnant women who are Rh negative. And flu shots with Thimerosal may be given to pregnant women. If a pregnant woman received all three of these injections, the harm to her baby would be absolute and permanent.

Hundreds of thousands of babies are born with blood mercury levels linked to IQ loss. These mercury-laced injections are lowering the IQ of babies according to a study by researchers at the Mount Sinai Center for Children's Health. Their research found that hundreds of thousands of babies are exposed to mercury prior to birth resulting in lower IQ's. Dr. Leonard Trasande, pediatrician and lead researcher, confirmed that between 316,588 and 637,233 infants are born in the U.S. every year with umbilical cord blood mercury levels linked to IQ loss. Then, after birth, the baby is still susceptible to being injected with vaccines containing Thimerosal.

If you are pregnant and want to protect your baby from mercury, talk to your physician and make it clear you do not want to be given any injections with Thimerosal. Ask your physician about alternatives to these injections if you are Rh negative or if you were to begin premature labor. Have notes placed in your medical chart so a stand-in for your physician will not aim a needle with mercury at you.

"I can state that the certainty of the science supporting mercury as a major cause of autism is probably more

overpowering than the science behind any other disease process that I studied dating back to medical school." Dr. David Ayoub stated in an interview with Independent Media TV. Dr. Ayoub is the author of the report, "Pregnancy and the Myth of Influenza Vaccination – Is it Safe, is it Effective, is it Necessary? What the CDC documents reveal."

After the Simpsonwood meeting pharmaceutical companies, especially Eli Lilly, contributed $873,000 to then Senate Majority Leader Bill Frist. Senator Frist created the "Eli Lilly Protection Act" that did not allow the public to sue Lilly for Thimerosal injuries. The Act was slipped in as a rider on a homeland security bill in 2002 and was passed. The Act was repealed in 2003. There have been at least ten different pieces of legislation shielding Lilly and vaccine makers from potential liability since the Simpsonwood meeting. There is abundant self-interest action occurring but no public interest action.

Several countries banned Thimerosal long ago. Russia banned any mercury in vaccines in 1980. Other countries that banned Thimerosal decades ago include Austria, Denmark, Great Britain, Japan, Norway and Sweden. These countries do not suffer from an autism epidemic.

States have begun to take matters into their own hands. In 2004 Iowa was the first state to pass legislation banning any mercury in vaccines. As of February 2007 the following states have banned mercury in vaccines: California, Deleware, Illionois, Iowa, Missouri, New York, and Washington.

There have been safe, non-mercury preservatives for vaccines for decades. Dr. Maurice Hilleman, one of the original creators of Merck's vaccine programs, was concerned about the potential harm to infants and children. He recommended decades ago that Merck discontinue using Thimerosal and replace it with available nontoxic preservatives. Merck did not take his advice based on a purely financial decision. It would cost more. They gave no regard to the safety concerns even though it was their own expert advising them to stop using Thimerosal.

A Grain of Sand Compared to Mount Everest

The increased costs to the vaccine makers compared to the past, present and future costs for the millions of children with mercury-caused neurological disorders including autism is like a grain of sand compared to Mount Everest.

Mark R. Geier, M.D., Ph.D. fought for years to have the CDC data released as the Freedom of Information Act allowed. Dr. Geier is eminently qualified to review the CDC data. He is board-certified in medical genetics and forensic medicine. He has studied vaccines for more than 30 years. His research and commitment to safety was key to convincing government officials to stop use of the whole cell Diptheria-Tetanus-Pertussis (DTP) vaccine. The change to the safer version without whole cells (DTaP) saves lives.

Dr. Geier was finally allowed in 2005 to study the Vaccine Safety Datalink (VSD) database of the CDC. He and his son, David Geier, are the only independent researchers who have been allowed access. Their analysis was published in the March 10, 2006 issue of *American Physicians and Surgeons*. The analysis shows that when children were given vaccines without Thimerosal the autism rate dropped significantly. Children and adults continue to be poisoned by mercury from pollution. Such as the air pollution from coal burning power plants putting 48 tons of mercury in the air every year in the U.S. But getting mercury out of vaccines has been statistically shown to decrease autism significantly.

Meanwhile the federal agencies that put great effort into hiding the truth are still denying the truth. As of March 2007 the FDA website for vaccines states, "The committee concluded that this body of evidence favors rejection of a causal relationship between Thimerosal-containing vaccines and autism." With lawsuits now coming to courts they do not dare change their statement and admit their disregard for guarding the public's health.

"Thimerasol used as a preservative in vaccines is directly related to the autism epidemic. This epidemic in all probability may have been prevented or curtailed had the FDA not been asleep at the switch." States the conclusion of the *Mercury in Medicine Report* published May 21, 2003 by the Subcommittee on Human Rights and Wellness of the Committee on Government Reform. The report went on to say, "The public health agencies' failure to act is indicative of institutional malfeasance for self protection and misplaced protectionism of the pharmaceutical industry."

Mad as a hatter. **Mercury is the second most toxic element on earth**, plutonium being the first. Without any research findings, what would your gut reaction be if you were asked: Do you think it would be safe to put the second most toxic element on earth in vaccines for infants and young children?

Albert Einstein said:
"The intuitive mind is a sacred gift
and the rational mind is a faithful servant.
We have created a society that honors the servant
and has forgotten the gift."

The decision makers of vaccines manufacturers and government officials in the agencies created to protect the public who have hidden and whitewashed the Thimerosal data have bright, rational minds. But show no signs of conscience.

One excuse used by some of these people is that they do not want the public to refuse having children immunized for childhood diseases. What is so terrible about a case of mumps or measles? The child is ill for a couple weeks. What is that compared to a life-long sentence of mental deficiency? People with autism and other neurological disorders due to mercury poisoning will always need supervision. Who will care for all the autistic and neurologically impaired people when their parents are too elderly or have died? Who will pay for the decades of care that will be needed? Costs the size of Mount Everest pouring down the mountainside towards every taxpayer.

Parents are being coerced into allowing vaccinations of their children. In Maryland parents on welfare lose $25 a month per child without immunizations. Many school systems require vaccinations before children can attend school. The schools do not realize that the requirement enlarges their own problem of the high incidence of special needs students.

It's time for parents to take back control of their children's wellbeing. Parents have the right to demand mercury-free vaccines only are given to their children. Parents have the right to say no to any vaccinations. Consider what disease the vaccine is for and just what are the chances your child might get the disease and how serious is the disease if the child did have it. Children may be far better off letting their own miraculous immune systems fight off an illness than be burdened for life by a debilitating neurological disorder and mental deficiency from a vaccine. A child who gets chickenpox has lifelong immunity and chickenpox is not serious for children. The vaccine does not offer lifelong immunity so people vaccinated in childhood can get chickenpox as an adult when it is much more serious.

Thimerosal is still used in influenza (flu), tetanus-diptheria and boosters, Hepatits, HIB, and meningitis vaccines. Flu vaccines without mercury can be requested. The old vaccines containing Thimerosal have a long shelf life. You have the right to ask and be shown the package insert of vaccines to ensure you or someone you love is not given a mercury-laced vaccine. You also have the right to say no to vaccines you cannot be certain are mercury free. Don't be bullied into an act that could cost your child his/her future. Even adults should be concerned about getting vaccinations containing mercury. Mercury damages the immune system. Do you want the second most toxic element on earth injected into you or your children?

The next time the government wants to immunize your child, think twice and ask questions. The Amish don't allow immunization of their babies and children. Amish children do not suffer from autism. And remember the U.S. infant mortality ranking in the world is 42nd. The U.S. does not have anywhere near the healthiest babies in the world.

Children are our most precious resource. They are the future. Millions of American families face a difficult and expensive future with autistic and neurologically disabled children. What have these people without conscience done to the future of millions of families, our nation, and the world?

12

THE PUPPET AND THE VENTRILOQUIST

Imagine a puppet sitting on the lap of a ventriloquist - an excellent ventriloquist. You find yourself watching the puppet's movements and listening to the puppet's jokes. Your attention stays on the puppet because a very good ventriloquist knows how to hold your attention there. Drug companies and their advertising agencies use the same act. The puppets convince the public common life conditions are actually drug treatable disorders. This is also known as disease mongering. It makes billions of dollars for the drug industry every year.

Sometimes the puppet is an admired movie star or a sports celebrity. Admire these folks for their acting capabilities or their sports ability but don't consider them astute medical professionals. They are simply being paid to use their celebrity status to help sell a drug. Don't be bamboozled by fame.

Laura Hutton had a contract with Wyeth to convince women menopause was not a natural condition. Instead menopause was labeled as a medical disorder of estrogen deficiency for which Wyeth had a drug. The advertising agencies understand women's desires. Ms. Hutton is attractive and admired. Women desire to be attractive and remain youthful. This person they admired was telling them there is a pill that would fulfill their desires.

Many alleged benefits of Wyeth's Premarin and Pempro were implied in the company's marketing campaigns to physicians and the public. The truth is the opposite. Instead of providing cardiovascular protection, Premarin and Pempro actually increase the risk for heart attack, stroke, blood clots and breast cancer. The increased cardiovascular risk had shown up in early clinical trials. Instead of further investigating this potentially fatal adverse event risk, Wyeth increased the noise level of marketing to create fear of menopause. The drug company continued to spread false claims of the benefits of its drugs. Instead of preventing Alzheimer's, another claim intimated by Wyeth, their drugs actually double the risk of women developing dementia.

Drug companies use medical conferences as their puppets. This was the case at the international menopause congress held in Sydney, Australia in 1996. The drug industry helped fund the congress. Drug companies, including Wyeth, paid for nearly half of the scientific sessions. Everyone's eyes were on the puppet. The science presented was biased, excluding negative outcomes.

Drug companies pay hundreds of millions of dollars to public relations (PR) agencies and advertising agencies every year to help them be excellent ventriloquists. These agencies create names for new health disorders from health conditions that are inconvenient. So constipation gets branded as "irritable bowel syndrome." Menopause is branded "estrogen deficiency." Emotional variations are branded as psychological disorders.

Many health experts not associated with pharmaceutical companies disagree that the conditions are disorders requiring drugs. The only purpose of these designer diseases is to create instant new markets for drugs; disease mongering. It is about money, not health. As reported in *Selling Sickness,* "the ability to create new disease markets is bringing untold billions in soaring drug sales." Why is the American public so quick to buy it?

After a new disease is created, drug companies finance the start-up of nonprofit patient advocacy groups, ostensibly to help "educate" the public about the new disease. If these groups suggest there are drugs that can help, you can be certain you are watching a puppet. Somewhere in the background the ventriloquist (drug company) is in control.

Patient advocacy groups and societies that accept money from drug companies "tend to advocate for faster review and availability of drugs, greater insurance coverage, and they tend to see direct to consumer advertising as a benefit to patients." Is the conclusion of Sharon Batt of Dalhousie University in Halifax, Canada. She has been studying advocacy groups for years, initially with breast cancer advocacy and recently with depression. She found groups maintaining independence from drug companies, receiving no money, in her words, "emphasize safety over speed and are critical of direct to consumer advertising."

Eli Lilly's use of the largest nonprofit drug puppet, the National Alliance for Mental Illness (NAMI), has led to racketeering charges. A lawsuit filed in New York in August 2004 accuses Lilly of violating racketeering laws, in part, by bankrolling NAMI to promote their antipsychotic drug Zyprexa for unapproved uses and to whitewash the drug's serious side effects. According to *New York Times'* investigative reporter Gardiner Harris, evidence to support the allegations also comes from Kentucky, where their Medicaid program was $230 million in the red in 2002. Zyprexa was the state's single largest drug expense and a state panel proposed, based on clinical studies showing the ineffectiveness of Zyprexa and the high risk of diabetes, to exclude Zyprexa from the Medicaid list. Kentucky held public hearings. Lilly paid for busloads of NAMI protesters to arrive at the hearing and also paid for full-page ads in Kentucky newspapers attacking the proposal under the guise of NAMI. Fortunately for Kentucky residents, Lilly lost the battle and Zyprexa was removed from the Medicaid list.

Residents in Oregon did not fare as well when the state decided to drop thousands of people from its Medicaid program in August 2003, rather than exclude ineffective, dangerous drugs like Zyprexa.

Zyprexa's serious side effects and Lilly's marketing of the drug for non-approved use, including use with children, has led to nine states filing lawsuits against Lilly as of May 2007 to recoup millions paid through Medicaid. Several lawsuits for the same reasons have also been filed in Canada.

Patient Education Puppets

Patient education packets and brochures about diseases are popular drug industry puppets. Wyeth used these puppets in Australia in 2000. The education packet warned women of possible health problems from menopause. These education packets failed to provide any true scientific data such as the well-designed HERS study run by the University of California, San Francisco that found Wyeth's Prempro actually increased the risk of heart attacks in women with heart disease. Women who read the packets were unknowingly watching the puppet and were not aware it was actually a ventriloquist lying to them.

Societies, advocacy groups and education packets are all methods drug companies use to get around the laws of countries that do not allow direct-to-consumer advertising by drug firms. These effective puppets are also used widely in the U.S. because they appear to be public-health-oriented and non-selling. The fact is they are totally contrived and controlled by pharmaceutical companies and their advertising partners. If any information eventually leads to a drug, you can be certain it started with drug firms. The ventriloquist is always about selling drugs no matter what the puppet appears to be about.

In April 2002 the Society for Women's Health Research (SWHR) held an expensive black-tie event for invited guests in Washington, D.C. Wyeth Pharmaceuticals was the "Grand Benefactor" for the elegant party with several other drug firms named as "Benefactors." One participant of the event said, "Without mentioning Wyeth it was like they were doing an ad for Wyeth." A week later SWHR received $250,000 from Wyeth at another special event. The HERS study results had been published four years earlier but were never mentioned at any puppet events.

Trying to hear the truth about drugs is like trying to hear a cricket during a rock concert. By 2002 the National Institute of Health had a decade of results from its Women's Health Initiative (WHI) study. This landmark study was the first large, long-term trial on healthy women. Women with uteruses were given Prempro, a combination of synthetic estrogen and progesterone. Women without uteruses were given Premarin, synthetic estrogen only. Both groups also had women taking placebo pills as control groups. NIH funded and ran the study that started in 1991 and was to have been a fifteen-year study. Wyeth provided the drugs. Are we to believe Wyeth was clueless to the adverse events study participants suffered? By July 2002, only three months after the Wyeth-sponsored Washington events promoting their hormone replacement drugs, the study was ended early because both Premarin and Prempro were doing more harm than good. The study groups taking the Wyeth synthetic hormones had significantly increased risk of heart attack, stroke, blood clots and breast cancer. Major media coverage presented the news as adverse events associated with hormone replacement therapy, not specifying it was just Wyeth's synthetic hormones that had been used in the study. The ventriloquist continues to try to discredit both the Women's Health Initiative study and the HERS study.

> Important warnings continued to be required on Wyeth's synthetic hormones:
> January 9, 2003: The FDA required warnings for increased risk of heart disease, heart attacks and breast cancer.
> February 2, 2004: The FDA required warnings for dementia and abnormal mammograms.
> July 29, 2005: The United Nations Cancer Agency reclassified synthetic hormone replacement as a carcinogen (causes cancer).

Synthetic estrogen was one of the leading fifteen drugs causing serious adverse events (side effects) reported to the FDA adverse reporting system between 1998 and 2005 with more than 11,500 reported events, a fraction of the actual.

This is one time when flashy, noisy puppets have not been successful in drowning out good science. Wyeth has seen sales of their synthetic hormones plummet from $2.07 billion in 2001 (before WHI results) to $880 million in 2004. In its latest attempt to slow the decline of sales, Wyeth filed a petition with the FDA in October 2005 to regulate natural (also called bioidentical) hormones originating from plants and formulated to individuals' needs at compound pharmacies. Many women and physicians have discovered the benefits of natural hormones without the concern for toxic side effects.

In the early twentieth century compound pharmacies served communities' healthcare needs by compounding to individual patient needs medicines from plant sources. Natural compounded medicines cannot be patented because they are from naturally available sources. As synthetic medications were manufactured in an industrial mass-produced mode by drug firms, the medical system began to change. Compounded medications were seen as old-fashioned. The ability to patent synthetic drugs and mass-produce them meant a new industry in post World War II U.S.

Now compound pharmacies are again recognized as important contributors to healthcare. Mainstream doctors as well as integrative physicians appreciate the advantages of natural medicinal agents with the dosage made specifically for each patient's needs. The FDA has received over 40,000 letters from physicians, women relying on natural hormones, and compound pharmacists in support of natural hormones since Wyeth filed the petition.

How pervasive is drug industry control of our healthcare choices? Two of the largest health insurance companies have recently changed their coverage of compounded natural hormones. In May 2007 BlueCross BlueShield stopped coverage of compounded medicines. Aetna Insurance announced discontinuation of coverage of compounded medications after October 2007. These companies will pay for prescriptions of synthetic hormones proven to cause heart attacks, strokes and cancer. All are serious and expensive to treat if the woman doesn't die. Expensive treatments drive up the cost of health insurance. The cost to women's health is immeasurable. Remember, 50 percent of women suffering their first heart attack die within one year.

The Society for Women's Health Research also received significant funding from drug maker Novartis so the society could provide an "educational" campaign to make women aware of another new disease condition, irritable bowel syndrome. The educational campaign included full-page national magazine ads and a web site. Women should see their doctor if they had any symptoms the puppet explained. Novartis had a pill of course, Zelnorm. Novartis sales representatives were pushing Zelnorm at physicians. As mentioned earlier Zelnorm was approved by the FDA in July 2002 and was withdrawn from the market on March 30, 2007 because of the risk of cardiovascular adverse events including heart attacks, strokes, and angina. But the puppet had done its job. Zelnorm made billions of dollars for Novartis in the almost five years it was on the market.

"Depression is Real" was a national campaign launched in 2006 by companies that make antidepressant drugs. The FDA has known for two decades the deadly dangers of antidepressant drugs in the classes of SSRI and SNRI. More about these deadly drugs is discussed in chapter fifteen.

'National Depression Screening Day' was created by drug companies and their advertising agencies to sell dangerous antidepressant drugs. They used numerous advocacy group puppets to promote the campaign including the: American Psychiatric Foundation, Depression and Bipolar Support Alliance, National Alliance of Mental Illness, League of United Latin American Citizens, National Medical Association, National Mental Health Association and the National Urban League. All of these organizations receive substantial funding from drug companies. The campaign included national television, radio and major city newspapers advertising, including *USA Today* and *The New York Times*, all paid for by drug firms that market psychiatric drugs.

Puppet organizations such as those listed above sometimes make false statements that mislead the public. It is easy to misinform the public by what may appear as an official source of information. If any advertising makes claims of huge numbers of people untreated for a disorder or health condition, realize the source is actually the ventriloquist drug companies behind the puppets. Drug companies are always trying to increase the number of people buying their drugs. It is always about money, not health. The principle cause of suicide in the U.S. is the antidepressant drugs that the campaign promotes.

Drug companies mounted the *Depression is Real* campaign because they fear sales of antidepressant drugs will decline since the FDA finally required a Black Box Warning for risk of suicide on all antidepressants. More physicians have finally realized they may be prescribing someone's death sentence.

The *Depression is Real* campaign is disease mongering. Drug companies want people to think the normal emotional flows of life can be fixed with a "happy pill". The campaign would have people believe they have a medical condition if they are not in a constant state of happiness, and their feelings of sadness, depression, or anxiety are due to brain chemical imbalance. Both are completely false. But the propaganda is distributed very successfully through drug advocacy group puppets. The roar of marketing drowns out the truth of science.

Dr. Fred Baughman, a neurologist and author of several books and papers on psychiatric drugs has been asking medical professionals for years to prove with any valid study that even one psychiatric disorder is due to a chemical imbalance with a confirming physical abnormality. No one has provided even a shred of evidence. Dr. Baughman says, **"We have a 'diseasing' of emotional and behavioral problems – of life problems, with never a mention that the causes can be found in every day life difficulties...the drugging psychiatry-pharmaceutical cartel is too anxious to 'disease' and 'disable' people with real life problems and the emotional symptoms they beget."**

In the U.S. there are measurable reasons for the majority of people to feel anxious or depressed. Yale University professor Jacob S. Hacker in his recently published book, *The Great Risk Shift*, describes the economic roller coaster that both educated and non-educated U.S. workers face today. The instability of income rose significantly for both groups since 1990. The number of months out of work has risen for both groups since 1990, leaving many unemployed longer than unemployment benefits cover. The risk of loss of health insurance rose since 1990. The risk of loss of retirement funds rose since 1990. Americans are losing their homes at record rates. Professor Hacker explains, "Prudent choices can reduce but not eliminate exposure to the growing level of economic risk."

For the first time since the Great Depression of the 1930's the majority of Americans are spending more than they earn. The 2007 federal poverty level for a family of four was $20,650 for all states except Alaska and Hawaii. More Americans now live below the poverty level at some time in their lives compared to the past three decades, as shown by the frightening truth that **more than half of all U.S. children spend at least one year in poverty by the age of 18.**

Drugs are not the answer to people facing these huge life problems. The way to address life problems is to first recognize what they are so they are not invisible ghosts haunting you. Then you can explore methods to reduce your risk. Understanding today's world of economic risks and having a personal plan to be prepared for the turbulence of change is like having a plan in case of a natural disaster. You don't know if it will happen to you but you have the peace of mind that if it does your plan will help. You feel empowered because you are prepared. You are not a helpless victim. Better to be saving your money every month as part of your plan than to be spending it on dangerous psychiatric drugs.

Being anxious about economic volatility is natural and addressable. It is not a health disorder requiring mind-altering drugs. Drug companies would have every aspect of life labeled a disorder in order to sell more drugs.

All the costs for these expensive drug industry puppets flow down hill to you even if you do not take a medication. For some people the costs become very great if they suffer an adverse event like a heart attack, stroke, cancer or suicide attempt. If they survive, they may have large medical bills and months of rehabilitation. If they do not survive, their families may have large medical bills, funeral expenses, and deep grief.

13

NO VALUE ADDED WHILE DRUG PRICES SKYROCKET

A Recurring Cycle of More Wealth

There cannot be absolute certainty of the safety of any drug for every person because the human form is so complex with infinite interactions within the body and with the external environment. And death is a part of the life process we should accept not fear. But the current drug-centered medical environment in the U.S. is not promoting good health and wellness. The drug-centered medical environment is out of balance from the reality of life and the reality of living forms.

The pharmaceutical industry consistently demonstrates its focus is money, ever more money. There are some honest, credible people within the pharmaceutical industry who would like to create good medicines. But the evidence of fraud, lies, and repression of the truth about dangerous adverse events demonstrates this is an industry that does not honor true science. It is an industry that promotes dishonesty with intent to harm.

Ten drug companies paid ten men almost $183 million in 2006. The unfathomable wealth drug companies make every year and pay their chief executives indicates why we cannot hope for change of these systems. The minimum annual compensation in 2006 for a pharmaceutical chief executive officer (CEO) was just under $5 million and the most was more than $36 million (total compensation according to the Securities and Exchange Commission). The average CEO annual compensation paid by the ten leading drug companies in 2006 was more than $18 million. That is compensation for just ten people in one year. There are many other executive positions in drug companies besides the CEO. And there are many more than ten pharmaceutical companies.

Every day drug companies push to make more money. As the drug coffers increase most people in the system become greedy for more money, even those people making unimaginable wealth. The system goes far beyond the drug firms. Stockholders want more. Drug lobbyists want more. Some politicians want more. Advertising agencies want more. Drug store chains want more. You get the picture. The larger the drug income pie looks, the larger slice everyone wants. It becomes a vicious cycle of greed. As all these parties want more the prices of drugs go higher. The push increases to expand the market with the illusion of health disorders. Like cancer, it just keeps growing and spreading without regard for what it is doing to the host system.

Unimaginable Markups

What products have more than a 500,000 percent markup? Prescription drugs. The cost of the active ingredients in 100 tablets of Xanax is $ .024. The consumer price in the U.S. for 100 tablets is $136.79. That is a 569,958 percent markup. Here are more examples:

Drug	Active Ingredient Cost	Consumer Price	Markup
Prozac	$.11	$247.47	224,973%
Zithromax	$18.78	$1,482.19	7,892%
Zocor	$8.63	$350.27	4,059%
Celebrex	$.60	$130.27	21,712%
Lipitor	$5.80	$272.37	4,696%

Claritin	$.71	$215.17	30,306%
Keflex	$1.88	$157.39	8,372%
Paxil	$7.60	$220.27	2,898%
Tenormin	$.13	$104.47	80,362%
Prilosec	$.52	$360.97	69,417%
Vasotec	$.20	$102.37	51,185%
Zoloft	$1.75	$206.87	11,821%

The above figures were compiled by Sharon Davis and Mary Palmer, Budget Analysts at the U.S. Department of Commerce. Generic drugs are also marked up as much as 3000 percent by some drug stores.

Of course there are costs besides the active ingredients. There are research costs but much of that is being moved offshore to cut the costs. Research costs represents only 14 percent of sales in the top ten companies. Patents on drugs are normally 20 years, allowing sufficient time to recoup research expenses.

Total real spending on drug promotions almost tripled between 1996 and 2005 to a whopping $30 billion per year.

"The apparent decline in FDA enforcement of direct-to-consumer drug advertising regulations calls into question the FDA's ability to prevent misleading messages about drug risks and benefits from reaching the public and heightens concerns about the potential adverse consequences such advertising might engender," Meredith Rosenthal, Ph.D. said in a news release. Dr. Rosenthal is associate professor of health economics and policy at the Harvard School of Public Health.

What value to the person taking a medication is added by the billions spent on drug promotions? How much value is added by the hundreds of millions of dollars paid to a few executives? How much value is added by the mob of drug salespeople invading physician offices? Is value added by the expensive puppets? Is value added by the millions of dollars paid to the 1,100 drug lobbyists in Washington? What value is added by millions of dollars paid annually as political contributions? Value isn't added by any of these costs but drug prices are continually increased to pay for them. The markup of pharmaceuticals in the U.S. will continue to rise as long as these expensive, non-value-adding practices continue.

With these extreme costs that do not add value to the medications it is no surprise that drug companies try to price fix and prevent free trade. Several drug firms now refuse to sell to Canadian pharmacies selling to American residents. Lobbyists have convinced members of Congress that public safety requires tougher regulations and enforcement of trade laws to prohibit Americans from purchasing their prescriptions from Canada or the United Kingdom. These are exactly the same pharmaceutical drugs from the same sources as the prescription drugs in the U.S. The only difference is the price. The extreme markup is not allowed in other countries. Each country has its own method to negotiate drug prices with drug companies, but the result is the same, no over-inflated drug prices for patients.

The pharmaceutical industry with their immense financial influence has won for now. On May 7, 2007 the Senate voted to continue restrictions on the importation of prescription drugs from Canada and other countries. That also means U.S. government agencies providing healthcare cannot save hundreds of millions of tax dollars by purchasing from these safe, reliable sources. The Medicare system, thanks to Medicare Part D being written by drug companies and their strong-armed techniques to force passage by Congress, is not allowed to negotiate for reasonable drug prices with drug firms or purchase from any other countries. Billions of tax dollars wasted every year.

States and cities suffering from deficits and concerned with the upward trend of healthcare costs for state employees and retirees are taking steps to provide reasonably priced drugs. Several states are directing residents to websites of foreign pharmacies they have prescreened for safety. These problem-solving governors and mayors have had to struggle with the federal government. For example, Illinois finally went ahead without FDA approval in late 2004 and launched Illinois' state website for prescription drug purchases. The website, www.I-SaveRx.net, allows people to purchase refills of prescription drugs from licensed, state-inspected and approved pharmacies in Canada, England and Ireland.

Drug companies don't want to lose their cash cow – the American public. British law does not allow drug companies to impose quotas on British pharmaceutical wholesalers. But within a week of Illinois' announcement of its website, Pfizer tried to impose a quota on British pharmacies to prevent sales to Americans.

Then there are the truly horrific health problems of some poor third world countries. In South Africa more than 30 percent of pregnant women were living with HIV in 2005. The estimate of HIV prevalence in the general population between ages fifteen and forty-nine years old was more than 16 percent in 2005 according to surveys conducted by the South African Department of Health. Under World Trade Organization (WTO) rules, a developing country has options to obtain needed medications under compulsory licensing or importation of generic, cheaper versions of the drugs even before patent expiration. South Africa created a Medicine Act, based on the WTO rule, to address its AIDS epidemic. The Act allowed generic AIDS drugs to be imported and produced in South Africa. In March 2001 forty-one pharmaceutical companies sued South Africa for its production of generic AIDS medicines. Worldwide protests against the drug companies caused the lawsuit to be dropped.

14

GOLD MINES

"There is no evidence that any mental disorder is caused by chemical imbalance," a Surgeon General's report states. But that truth is not slowing the sharp increase in the prescriptions written for psychiatric drugs.

In 2007 the five leading psychiatric drugs grossed more money than the gross national product of half the countries in the world.

The pharmaceutical industry has discovered two gold mines that could potentially keep billions of U.S. tax dollars flowing into their coffers every year. One gold mine is their ability to influence government legislation requiring mandatory mental health testing. The legislation has the potential of leading to millions of prescriptions annually. The other gold mine is the industry's influence over the creation of drug guidelines, also called algorithms, defining what psychiatric drugs will be used in large public systems like state mental hospitals, prisons, and clinics funded by state and federal governments.

Many psychiatrists are alarmed by the influence the drug companies have on the mental health field. **"There is not one shred of credible evidence that any respectable scientist would consider valid demonstrating that anything that psychiatrists call mental illness are brain diseases or biochemical imbalances. It's all fraud,"** states Dr. Ron Leifer a psychiatrist.

Every mental disorder in the U.S. is created BY VOTE by the American Psychiatric Association (APA). There are no diagnostic tests to provide evidence that mental health disorders are caused by biochemical or neurological conditions. It is a myth that psychiatry and the drug industry initiated. The names of the disorders and the symptoms are all voted into existence! There is no valid test - no blood test, no brain scan, no tissue test, no hormone test, no MRI, nothing at all that can be used to diagnose any of the more than 350 mental disorders now listed in the *Diagnostic and Statistical Manual for Mental Disorders (DSM)*. And the APA continues to create new disorders and new symptoms at an alarming rate. A longer list serves both psychiatry and the drug industry. One English psychiatrist candidly states in the documentary film, *Where the Truth Lies*, "DSM stands for diagnosis as a source of money. It brings in a lot of money." Psychiatrists from around the world were included in the documentary, and every one of them acknowledged that **the DSM is not backed by science.** Which means that no mental disorder is backed by science. If you want to know the interesting history of the myth of brain chemical imbalance, go to http://thehealthyskeptic.org/the-chemical-imbalance-myth/.

> "Those who can make you believe absurdities
> can make you commit atrocities."
> Voltaire

The Pharmaceutical Industry
and the Psychiatric Establishment
Are Conjoined Twins – Joined at the Wallet

The conjoined twins have been able to make millions of people, including physicians, believe in absurdities, which has led to people committing atrocities such as drugging children with deadly psychiatric drugs.

The APA is the most drug industry financially supported medical association. In July 2008 Senator Charles Grassley's demands that the APA provide an accounting of its finances uncovered that in 2006 the drug industry accounted for about 30 percent of the APA's financing; more than $20 million dollars.

Senator Grassley's investigations have also disclosed drug industry payments to individual psychiatrists. The APA president, Dr. Alan F. Schatzberg of Stanford University has $4.8 million stock holdings in a drug development company.

Psychiatrist Dr. Melissa P. DelBello of the University of Cincinnati reported working for eight drug companies. She claimed her earnings from drug makers totaled about $100,000 from 2005 to 2007. But one drug maker disclosed to Senator Grassley that the company had paid Dr. DelBello more than $238,000 in those two years.

Two child psychiatrists from Harvard Medical School, Dr. Joseph Biederman and Dr. Timothy E. Wilens, reported to the senator that they each earned several hundred thousand dollars from drug companies from 2000 to 2007. However it was found they each earned at least $1.6 million.

CONJOINED TWINS OF PHARMA AND PSYCHIATRY

"In order to survive we psychiatrists must go where the money is," Dr. Steven Sharfstein, APA vice president told Congress. The money is in drugs.

There Is No Separation of Body-Mind-Spirit

Physicians, nurses and other healthcare workers have warned that symptoms of voted-in mental illnesses can be caused by genuine physical illnesses and treated medically, not with psychiatric drugs. A good example is hypoglycemia, which is low blood sugar. Symptoms of hypoglycemia include fear, anxiety, disorientation, a sensation the body is going to die, sweating, clammy hands, and acting erratically. All are voted-in symptoms of several created-by-vote psychiatric disorders. Hypoglycemia can be confirmed with blood tests and a person can usually control it with modified eating habits and no drugs.

Another example is heart valve prolapse. It causes the sensation of rapid heartbeat, fluttering sensation in the chest, sweating, and anxiety. These are also symptoms defined by the APA to indicate panic attacks. But anti-anxiety drugs do not help. Heart valve prolapse requires medical care by a cardiologist.

Food allergies, other types of allergies, and dehydration can also cause symptoms listed for mental health disorders. In fact, symptoms of most, if not all, physical medical conditions can be found as symptoms of APA voted-in mental disorders. The body-mind-spirit functions as an integrated system. There is great harm done every year to hundreds of thousands of people because our conventional medical system does not appreciate whole-person healing. And the conjoined twins of psychiatry and the pharmaceutical industry take full advantage of the archaic mode of compartmentalizing the human system.

A Tale of Digging for Gold in Texas

In 1995 an alliance formed. Participants from the pharmaceutical industry, the Texas state university and Texas mental health and corrections systems proclaimed to have a goal of developing a model mental health treatment program. The true objective of the alliance was to open public institutions to the drug companies' control, ensuring profits of billions of tax dollars annually.

The funds for the group included a $1.7 million grant from a Johnson & Johnson-related foundation. J&J owns Janssen Pharmaceuticals, the maker of the atypical antipsychotic Risperdal. The aim of the pharmaceutical companies was to get their new, more expensive psychiatric drugs named as the first to be used on every patient. The alliance created a menu of drugs dictating the order in which they were to be given for various psychiatric conditions. They called it the Texas Medication Algorithm Project (TMAP).

Robert Whitaker, author of *"Mad in America"* and an award winning science journalist, found drug companies had made false claims about the new antipsychotic drugs for schizophrenia, referred to as atypicals. Whitaker learned the FDA's review of clinical trials did not support the drug companies' claims that atypicals were safer and more effective than existing generic drugs. The FDA approval letter to Janssen for Risperdal stated:

"We would consider any advertisement or promotion labeling Risperdal false, misleading or lacking fair balance under section 502(a) and 502(n) of the ACT if there is a presentation of data that conveys the impression that Risperdal is superior to haloperidol (*a generic*) or any other marketed antipsychotic drug product with regard to safety or effectiveness."

But the Texas alliance, with its financial backing from the makers of atypical antipsychotic drugs, wanted to put new expensive drugs, Risperdal and Zyprexa, as TMAP's leading choices for physicians in Texas facilities to use on all patients. The alliance invented a way to legitimize the medications they recommended. They selected 57 psychiatrists and experts to take a survey the alliance had designed about certain drugs. Then the alliance analyzed the surveys and made conclusions. They called the process "Expert Consensus Guidelines." The Expert Consensus did not include review of clinical trials, any studies, or analysis of costs. In other words, there was no objective science used by the Expert Consensus. Most of the experts selected by the alliance had strong financial ties to the drug industry.

The results of the Expert Consensus Guidelines were:

1. Risperdal (Janssen Pharmaceuticals)
2. Zyprexa (Eli Lilly)
3. Seroqual (AstraZeneca)

for treating schizophrenia, in that order. Also, without the use of objective science, the Expert Consensus Guidelines specified the newer, brand name anti-depressants were superior to generics and the newer, branded drugs for bi-polar were superior to generics. The alliance stated their Expert Consensus had established all these drugs were safer, more effective, better tolerated and relatively free of side effects when compared to the older, generic drugs. This became the TMAP in 1996.

As of July 2008, ten states have brought lawsuits against Eli Lilly for fraudulently marketing Zyprexa. Canadian provinces have also filed lawsuits against Eli Lilly for fraudulently marketing Zyprexa.

The Texas Attorney General in 2007 sued Janssen Pharmaceuticals for the fraudulent marketing of Risperdal for off-label uses that cost the state's Medicaid millions of dollars.

TMAP was fully supported by then-Governor George Bush and the Texas Legislature. They funded it and opened the doors of state facilities to the guidelines and the state's checkbook to the drug companies.

Physicians working in any of the state-run institutions were required to use TMAP for all patients. If doctors wanted to use a generic drug, they had to write a report rationalizing their decision. Patients in the system were switched to the newer drugs without any medical indication to do so. Many patients had been taking generic drugs for years. If the physicians didn't like it, they would have to look for new jobs. Medical practice by administration is not focused on the health needs of patients. TMAP also did not save taxpayers' money. The TMAP-required drugs were all much more expensive than the older generic drugs.

Psychiatrists Speak Out Against the Trend

In 1999 one of the experts spoke out against the Expert Consensus method. "The most important weakness of the Expert Consensus Guidelines is that the recommendations are based on opinions, not data. History shows that experts' opinions about best treatments have frequently been disproved, and there is no assurance that what the experts recommend is actually the best treatment. One danger here is that clinicians or administrators may misinterpret current consensus as truth." Peter J. Weidman, M.D. wrote in his article, *Guidelines for Schizophrenia: Consensus or Confusion?* The article was published in the *Journal of Practical Psychiatry and Behavioral Health* in January 1999.

Dr. Weidman was one of the experts on the Texas Expert Consensus. He explained in his article that the treatment options on the survey given to the experts were too limited to cover real life situations. He identified the bias of the survey created by the funding source, Janssen Pharmaceuticals, maker of Risperdal. He also confirmed the conflict of interest because most of the experts selected had financial ties to the pharmaceutical companies that made the psychiatric drugs being considered.

"Our (*psychiatry*) field as a whole is progressively being purchased lock, stock and barrel by the drug companies. This includes the diagnoses, the treatment guidelines, and the national meetings." Daniel J. Carlat, M.D. told *The Boston Globe*. Dr. Carlat is a psychiatrist and former paid speaker for drug companies. He realized one day, "I was being paid to say good things about drugs, regardless of what my actual opinions were." He stopped working with drug companies and now publishes *The Carlat Psychiatry Report*, a monthly newsletter available on the Internet.

"Psychiatrists are among the most conflicted of the medical specialties." According to Dr. Jerome P. Kassirer, a Tufts University professor and author of the book "*On the Take.*"

"The very vocabulary of psychiatry is now defined at all levels by the pharmaceutical industry." Reports Dr. Irwin Savodnik, a University of California Los Angeles (UCLA) psychiatry professor.

The data certainly supports these physicians' statements. One hundred percent of the psychiatric panel members determining the symptom guidelines for 'schizophrenia', 'mood disorders' and 'other psychotic disorders' have financial ties to the drug industry, as disclosed in "Financial Ties Between DSM-IV Panel Members and the Pharmaceutical Industry" published in *Psychotherapy and Psychosomatics* in 2006.

The New York Times found psychiatrists were paid more by drug companies than any other medical specialty in Minnesota, the only state requiring full reporting of financial relationships between physicians and drug companies. Between 1997 and 2005 psychiatrists in Minnesota collected $6.7 million from drug companies. *The New York Times* also found drug companies were not selective about psychiatrists paid to conduct clinical trials of drugs. At least 103 doctors had been disciplined, criticized, or had their license to practice revoked by the Minnesota state medical board. In the case of one psychiatrist, the FDA concluded he had violated the protocols of every drug study he led that the FDA had audited. The FDA found he had reported inaccurate data to the drug makers. Despite all this, drug makers continue to hire this psychiatrist. It appears quality, truth, and good science are not required in psychiatric drug testing.

Scientific Evidence of TMAP Psychotic Drugs

One in every 145 patients died in clinical trials of those taking the atypical drugs Zyprexa, Risperdal, or Seroqual for schizophrenia according to FDA data. The deaths are not mentioned in any scientific literature. The drug companies did not want physicians and the public to know. These were the three atypical drugs of choice on the TMAP.

In the Zyprexa clinical trials 22 percent of the patients taking the drug suffered serious adverse events. That means, nearly a quarter of the people taking Zyprexa had a life threatening event or required hospitalization from taking the drug, according to the FDA's definition of serious adverse event. One in every 35 patients taking Risperdal in the clinical trials experienced a serious adverse event.

"There is no clear evidence that atypical antipsychotic drugs are more effective or are better tolerated than conventional antipsychotic drugs. Conventional drugs should remain the first treatment..." was the conclusion published in 2000 in the *British Medical Journal* by Dr. John Geddes. His study reviewed 52 clinical trials. More than 12,000 patients were in the trials. The British Department of Health funded the study with no funding from pharmaceutical companies.

The National Institute of Mental Health (NIMH) conducted their own study in an effort to determine whether atypical antipsychotic drugs were worth their enormous cost. They designed an 18-month study. NIMH spent $44 million on the study and reported in September 2005 the same conclusion that Dr. John Geddes had reached in 2000. The new atypical drugs "have no substantial advantage" over the older, much less costly drugs.

Drug company influence over government decisions appears to be as strong at the federal level as it was in Texas. With all the scientific evidence of the extreme dangers of these drugs, in October 2006 the FDA approved Risperdal for use in children as young as 5 years old to treat irritability of autistic children. Approval was given based on just two clinical trials each lasting only eight weeks. The FDA website indicates that 100 percent of the 40 children taking the drug in one trial suffered adverse events after taking the drug for only eight weeks. In real-life situations children may be placed on this dangerous drug for years, if the children live. The approval was given with no disclosure data for independent review, no advisory committee, and no public hearing, according to the Alliance for Human Research Protection.

The FDA gave another Risperdal approval on August 22, 2007 for children as young as 10 years old for "short-term treatment of manic or mixed episodes of bipolar 1 disorder." The FDA news release reports their approval was based on Janssen's two short clinical trials lasting only 6 and 8 weeks.

This is the same drug that caused one in every 35 adults in clinical trials to suffer a life-threatening event or required hospitalization. Adults have died from this drug. Adverse events suffered by adults from Risperdal reported to the FDA include: sudden death, cardiopulmonary arrest, anaphylactic reaction, angioedema, apnea, atrial fibrillation, cerebrovascular accident, diabetes, hyperglycemia, intestinal obstruction, mania, and pulmonary embolism.

An astounding $14.6 billion was reaped by U.S. sales of atypical antipsychotic drugs in 2008. Much of that money came from tax dollars through Medicare, Medicaid and the Veterans Administration. Soldiers returning from the Mideast are often prescribed these drugs, with deadly consequences.

Dr. Geddes' thorough review of 52 clinical trials was published in 2000. But U.S. federal and state agencies continued to spend billions of tax dollars for years and are still spending on these drugs in 2009. This is the waste of public money that is allowed to continue because the drug industry has purchased influence at all levels of government.

As the state lawsuits against the drug companies demonstrate, much of the public money spent is for illegal off-label (non-approved) use of these dangerous drugs. There are not the numbers of people suffering from schizophrenia or bipolar, the FDA-approved uses, to reap the billions of dollars annually. However, the conjoined twins of the drug industry and APA are doing their best to create a larger lasso to rope more people, adults and children, in for branding with mental disorders that lead to these drugs. And the FDA is helping the drug companies by approving these dangerous drugs for more uses in both adults and children. The question is: for the benefit of the public or the benefit of drug companies?

How ironic that the federal government does not allow U.S. citizens to legally purchase necessary life-sustaining medicines from Canada or Great Britain for reasonable prices, while the U.S. government spends billions of tax dollars every year for psychiatric drugs that have been repeatedly shown to have no benefit over the older, less expensive drugs, and have severe life-threatening side effects that will cost taxpayers billions of dollars more.

Zyprexa is among "the most deadly drugs ever to gain FDA approval." independent researcher and psychiatrist Dr. David Healey concluded after he reviewed FDA data of Zyprexa. Dr. Healey was the researcher who discovered Eli Lilly had hidden data about suicides and suicide ideation occurring in trials prior to FDA approval of Lilly's antidepressant drug Prozac. Prozac is a drug in the guidelines for children created by the Texas alliance.

Eli Lilly has paid at least $1.2 billion to 28,500 people who claimed they were injured by Zyprexa, reports Alex Berenson in *The New York Times*, January 2007. Lilly said the settlement did not change its view that Zyprexa is a safe and effective treatment. Lilly internal documents show they knew the risk of diabetes and obesity from clinical trials before the drug was approved.

Zyprexa global sales in 2006 were $4.36 billion, most of it in the U.S. Zyprexa remains on TMAP. Marketing continues to win over true science. Why aren't physicians listening to science that is not sponsored by drug companies?

TMAP Breaks the Bank

TMAP was initiated in 1996. An article in the June 18, 1998 edition of the *Abilene Reporter News* reported the Texas Mental Health and Mental Retardation (MHMR) agency was in severe financial trouble due to medication costs. The article by Jerry Daniel Reed described the system as "choking on the costs of new-generation medications that treat schizophrenia, depression and bi-polar disorder." A state official quoted in the article said, "I believe that our (mental health) centers are in crisis right now because they're trying to squeeze money out for these new medications." Medications that research had shown were not more effective or safer than the older, less costly drugs.

By February 2001 TMAP and TCMAP (the children's version) had bankrupted the Texas Medicaid program and the budgets of the state's mental health and prison system. Reporter Nancy San Martin wrote in the *Dallas Morning News*, "Texas now spends more money on medication to treat mental illness for low-income residents than any other type of prescription drug. Prescription drugs are the fastest growing expense within the healthcare system. And the cost for mental disorder treatments is rising faster than any other type of prescription drug."

Score: Texas Taxpayers – Zero, Drug Companies – Won

The same pattern is now happening at the federal level. The Veterans Administration (VA) spent $208.5 million just on atypical psychotropic drugs in 2003; $106.6 million just for Zyprexa. Zyprexa costs $3,000 to $9,000 more per patient annually than Haldol, the generic that clinical trials show is as effective and is much safer than Zyprexa. In the VA system 80 percent of people labeled with schizophrenia are currently treated with the expensive atypical antipsychotic drugs.

Score: U.S. Taxpayers – Zero, Drug Companies – Won

Taxpayers are not the only ones losing to the ubiquitous plague of the drug industry's powerful influence. People put on psychiatric drugs can lose the most. The *Journal of the American Medical Association (JAMA)* published results of a study finding **elderly people had a 54 percent increased risk of death** within

3 months when they were given atypical antipsychotic drugs for dementia, an off-label use of the drugs.

Nobel Prize laureate John Nash, Ph.D. had suffered from schizophrenia. His story was portrayed in the movie "A Beautiful Mind." But the movie erroneously portrays Nash on medications just before receiving the Nobel Prize. Sylvia Nasar, the author of Nash's biography from which the movie is based, identifies in her book that Dr. Nash stopped taking any antipsychotic drugs in 1970 and spent twenty years slowly recovering without drugs. Nasar concluded Dr. Nash's refusal to take drugs was what allowed him to recover fully. The drugs' deleterious effects "would have made his gentle reentry into the world of mathematics a near impossibility."

Nash's recovery without drugs is not unique. Dr. Loren Mosher, a Harvard trained physician, created the Soteria Project in 1971. The project provided a residence emphasizing empathy and care without drugs or psychiatrists for people suffering from schizophrenia. Soteria was a complete success. Residents became contributing non-psychotic members of society. But when Mosher began to publish the outcomes, the National Institute of Mental Health (NIMH), with its close ties to the pharmaceutical industry and the APA, denounced the credibility of the pilot project and ceased funding.

Ironically, European countries were inspired by the project. Switzerland, Sweden and Finland all have Soteria homes and report good results. Swiss researchers documented in 1992, "Patients who received no or very low-dosage medication demonstrated significantly better results."

Psychologist Courtenay Harding, Ph.D. found the majority of people diagnosed with schizophrenia do completely recover or significantly improve and function in the community if they are provided a rehabilitation support system. The majority of people supported with a recovery program are able to discontinue use of any medications.

People treated by psychiatrists with the mindset that schizophrenia is a life-long disorder requiring a life of drugs are at a disadvantage. The mindset creates a close-minded approach to providing healthcare.

People suffering from schizophrenia in poor countries have a much greater chance of recovery than people with schizophrenia in the U.S. and other drug-centered nations, was the surprising result of an eight-year study by the World Health Organization (WHO). Living in a developed nation was a "strong predictor" a person with schizophrenia would never recover, was the conclusion of the WHO researchers. As with Dr. Mosher's results, the U.S. psychiatric and pharmaceutical leaders simply dismissed the WHO study as flawed. The WHO study results were published in 1979 and three decades later the U.S. is still locked into a drug-centered mindset because of the powerful financial influence of the pharmaceutical industry.

Putting Children on Psychiatric Drugs

With their great financial win of millions of Texas tax dollars every year, the drug companies lost no time in pursuing additional sources of tax money. In 1997 the Texas alliance, funded by psychiatric drug makers, began to create the Texas Children's Medication Algorithm Project (TCMAP). Again no actual science was considered. Another expert consensus panel was assembled to determine the drugs that would be best for treating children and teenagers with mental or emotional problems. What these "experts" label as emotional problems are actually the natural process of development of children in a highly stressed society. Strongly varying emotions are part of the natural process of childhood and teenage development, making them an easy target for the conjoined twins of psychiatry and drug companies.

The Texas alliance assembled primarily the same "experts", most with financial ties to drug companies, who had been involved in the TMAP. They met and decided the drugs defined on the TMAP should also be used on children and teenagers. Many of the drugs were not FDA approved for children and adolescents.

While TCMAP panel members were speaking and writing about the safety and effectiveness of the antidepressant drugs they had put on TCMAP, two of the drugs, **Paxil and Effexor, were banned from ever being given to children in Britain** because of the high risk of suicide. Paxil is an antidepressant in a class of drugs known as selective serotonin reuptake inhibitors (SSRI). Effexor is an antidepressant in a class of drugs known as selective serotonin and norepinephrine reuptake inhibitors (SNRI).

Serotonin and norepinephrine are natural neurotransmitters in our bodies. They facilitate communication within the body and create a sense of well-being. SSRIs and SNRIs are meant to prevent the body's natural process of reabsorption of these natural chemicals in an attempt to keep them in the body longer to help the good feelings last longer. But the complex functioning of the human form isn't that simple. The body-mind is made to keep the hundreds of the body's natural chemicals in balance. That is why the body has the means to reabsorb natural chemicals so they do not accumulate in the body and put the intricate functioning out of balance.

In July 2006 the FDA issued an alert for "potentially life-threatening serotonin syndrome" as a side effect of these drugs. The FDA described the reported symptoms:

"The reported signs and symptoms of serotonin syndrome were highly variable and included respiratory failure, coma, mania, hallucinations, confusion, dizziness, hyperthermia, hypertension (high blood pressure), sweating, trembling, weakness, and ataxia."

Ataxia is unsteady and clumsy motion of the limbs or torso that occurs when parts of the nervous system controlling movement are damaged. The onset of the symptoms after patients took the drugs was as quick as ten minutes with the longest onset only six days.

Prozac was the first SSRI antidepressant and marketed as a wonder drug. The real wonder is how it received FDA approval and remained on the market with its deadly record.

15

ANTIDEPRESSANTS CREATE KILLERS

"The suicide rate is 718 for every 100,000 people taking SSRI/SNRI drugs in clinical trials." Dr. Arif Khan told a meeting of the New Clinical Drug Evaluation Unit of the NIH in August 2002. "We have to ask if medication is the only way to approach the prevention of suicide." Dr. Kahn had reviewed all the clinical drug trial data for SSRI and SNRI drugs approved by the FDA between 1985 and 2000. **The suicide rate for all ages in the general public is about 11 for every 100,000 not taking antidepressants.** "The high rate of suicide among patients who tested the drugs might suggest an 'iceberg effect' in the general population," Dr. Kahn told the NIH meeting. The people committing suicide in trials are only the tip of this horrific iceberg as millions of people are given these deadly drugs.

"Antidepressants can cause suicides." In 2003 British Medicines and Healthcare Products Regulatory Agency (British equivalent of the FDA) sent a letter to all British physicians and other health professionals giving stern warnings not to use any SSRI or SNRI antidepressants on anyone under the age of 18 years. Their review of data found the questionable benefits did not outweigh the potential risk of suicide with these drugs.

In November 2004, two years after Dr. Kahn disclosed his findings and seventeen years after the first SSRI had been approved, the FDA finally issued its first warning that SSRI and SNRI drugs increase risk of suicide in youths 18-years old and younger. A Black Box Warning on the package insert was required stating these drugs can produce severe adverse events (side effects), including: "anxiety, agitation, akathisia (extreme inner aggravation), panic attacks, irritability, hostility, aggressiveness, impulsivity, and mania." Mania is an extremely elevated mood that can cause unusual thought patterns, delusions and hallucinations. The FDA also warned that when people *stop* taking antidepressants it can cause suicide, psychosis, or hostility. But the FDA warnings are not sent in a letter or email to physicians. It is easy for new warnings printed on the package inserts (PI) to go unnoticed by physicians. Patients are often not told of any warnings.

The pediatric psychiatrist who put 12-year-old Candace (name withheld by family request) on Zoloft simply because she became anxious during school exams told her parents that the drug was safe and that Candace would be "happier" on the drug. Candace was on the drug only months when she hung herself in her bedroom in 2004.

Aaron Todovich was only 15 years old when a pediatric psychiatrist diagnosed him as depressed after a ten-minute conversation and put him on an antidepressant. Over the next ten years four psychiatrists prescribed several antidepressants, atypical antipsychotic drugs, and Ritalin. Why? Because the drugs caused side effects, and the psychiatrists' response was always to diagnose a new disorder and write a prescription for yet another dangerous, mind-altering drug. Aaron tried stopping the drugs several times. But no one told him or his mother that stopping cold turkey brought on severe reactions. No psychiatrist tried to help him safely stop the drugs.

As soon as someone is labeled a "patient" the attitude is that the person needs to be fixed. Whatever occurs is due to the imperfection of the patient, not the treatments. Effexor was the last antidepressant Aaron took. He ended up in the hospital with liver dysfunction, enlarged heart, and anemia. All are stated side effects of the drug, but the psychiatrist giving Aaron the prescription never spent any time explaining these life threatening side effects, or the depression that antidepressants can cause. Aaron committed suicide in November 2003. Finally he was free of the drugs and the psychiatric drug pushers.

Shawna Scantlin was 47 years old in 2006 when she committed suicide after her family physician put her on a dangerous cocktail of drugs including the highly addictive painkiller OxyContin and the addictive antidepressant Zoloft. When her "depression" about her chronic neck pain continued (because the OxyContin was not working), her family physician simply increased her daily dosage of Zoloft.

Drug companies try to confuse the public with media messages from paid experts. Dr. Charles Nemeroff, a psychiatrist who has financial ties with seven drug companies that market antidepressants, was reported as blaming the Black Box Warning for the increased suicide rate of adolescents.

But his claim is putting the cart before the horse. Before the FDA requires new warnings for any drug they have reviewed historical data. The Black Box Warning was finally put on when FDA leaders could no longer cover up the direct relationship between antidepressants and suicides.

Years of Cover-Up by Drug Companies and the FDA

Drug companies and the FDA have been aware of the high suicide and violence rates since the earliest antidepressant trials with the first SSRI drug, Prozac.

"During the treatment with the preparation (Prozac), 16 suicide attempts were made, 2 of these with success. As patients with a risk of suicide were excluded from the studies, it is probable that this high proportion can be attributed to an action of the preparation (Prozac)." A 1984 letter from the German pharmaceutical licensing authority, BGA, to Eli Lilly clearly states the BGA's concerns for the risk of suicide caused by Prozac.

Internal memos between Eli Lilly employees show Lilly knew from their own clinical trials Prozac (fluoxetine) greatly increased suicide and suicide ideation (thoughts of suicide).

Aug. 3, 1990: Memo to sales representatives regarding reports of suicidal ideation/behavior associated with Prozac therapy: "This information is not intended to replace our current promotional strategy but is being provided to enable you to respond to physicians when appropriate. You should not initiate discussion on these issues nor use this letter in detailing (*speaking with doctors*). However, if asked to comment on these issues by a healthcare professional you should reassure the healthcare professional that no causal relationship has been established between suicidal ideation and Prozac therapy."

Oct. 2, 1990: Memo between two Lilly employees regarding an upcoming Prozac symposium in which the issue of suicidality will be discussed: "Then the question is what to do with the big numbers on suicidality."

Nov. 13, 1990: Memo from a Lilly employee in Germany to an employee in the U.S. responding to Lilly's request to change words used for adverse drug event reporting about Prozac suicides. U.S. Lilly wanted to change "suicidal ideation" to "depression." German employee writes: "Hans (*another Lilly employee in Germany*) has medical problems with these directions and I have great concerns about it. I do not think I could explain to the BGA (*the German drug licensing authority*), a judge, a reporter, or even to my family why we would do this, especially on the sensitive issue of suicide and suicidal ideation."

One of Lilly's chief scientists wrote in an internal memo, "Anything that happens in the U.K. can threaten this drug

(Prozac) in the U.S. and worldwide ...We are now expending enormous efforts fending off attacks because of

1) relationship to murders and 2) inducing suicidal ideation."

'Even if we got several hundred reports involving suicide and Prozac, we wouldn't be alarmed.' Dr. Paul Leber, a psychiatrist and director at the FDA, told *Time Magazine* in 1990. Dr. Leber and other FDA leaders continued to help Lilly cover-up the extreme suicide danger of Prozac as Americans and people around the world continued to die.

While Dr. Leber was protecting Eli Lilly with his public statements, an FDA researcher was warning of Prozac's dangers in his internal FDA Prozac safety review. David J. Graham, MD, MPH, an FDA researcher, stated in his 1990 safety review:

> "The data showed higher percentages of suicidality among fluoxetine (*Prozac*) patients than among tricyclic (*old generation antidepressant*) or placebo patients...apparent large scale underreporting (*by Lilly*), the firm's analysis cannot be considered as proving that fluoxetine and violent behavior are unrelated."

Did Dr. Graham's safety warning concern FDA leaders?

"In summary, I don't consider these data to represent a signal of risk for suicidality for either adults or children," the FDA Team Leader of Psychiatric Drug Products, Dr. Thomas P. Laughren wrote in 1996, six years after Dr. Graham's analysis and warning, and thousands of deaths later.

The FDA's refusal to require the withdrawal of Prozac from the market left physicians unaware that they were writing prescriptions for death. Most physicians would have assumed patients' suicides were from depression, not the drug, which is why the number of suicides reported to the FDA is probably a small percent of the actual.

Black Box Warning For Suicide
On All 33 Antidepressants

In May 2007 the FDA finally required a new Black Box Warning for all antidepressants for increased risk of suicide in people under the age of 25. Is suicide a risk only for young people on these drugs? Not according to accumulating data. Seniors who were taking SSRI or SNRI antidepressants were nearly 5 times more likely to commit suicide in the first month on the drugs than seniors taking the older generation medications for depression, according to a recent study by Toronto's Institute for Clinical Evaluative Sciences. The study was reported in the *London Free Press* on May 1, 2006.

The list of these deadly antidepressants as of May 2007 has increased to 33 brands (see Appendix A for the complete list). The new Black Box Warning should be an imposing caveat to physicians. But will the message be heard and taken seriously over the loud noise of pharmaceutical marketing?

"An immense number of patients get worse from SSRI/SNRI antidepressant drugs," the Vice President of the Swedish Psychiatric Association, Dr. Christina Spjut, announced on Swedish national television on November 16, 2006. Sweden has mandatory drug adverse event (side effects) reporting. More than half, **52 percent, of all women in Sweden who committed suicide in 2006 had filled a prescription for an antidepressant within 180 days of their suicide.**

"The bottom line is that we really don't have any good evidence that these drugs work," states Joanne Moncrieff, M.D. professor of medicine at University College of London in an article published in the *British Medical Journal.*

The benefits of SSRI and SNRI drugs have repeatedly been shown in clinical trials to be minimal or nonexistent. FDA's own records show the agency knew Celexa was ineffective when the FDA approved the drug. The approval was given on just two marginally positive trials out of the 17 trials conducted. Celexa causes all the dangerous side effects, including suicide and violence.

Eli Lilly's analysis of Prozac in 1985 showed the drug failed to demonstrate efficacy in clinical trials. A March 29, 1985 internal Lilly document states: "The benefits vs. risks considerations for fluoxetine (Prozac) currently do not fall clearly in favor of the benefits." Patients in the trial taking the older tricyclic drugs or the placebo were better off than the patients who took Prozac.

Dr. Peter Breggin, psychiatrist and author of several books including, "*Talking Back To Prozac,*" reviewed the earliest clinical trials of Prozac and found they showed the drug was not effective and had a high risk of suicide and violence. "When this potential economic disaster for Eli Lilly was discovered, the FDA allowed the company to include in its efficacy data those patients who had been illegally treated with…tranquilizers in order to calm their over-stimulation (*from Prozac*)." Dr. Breggin clarified, "Basically, Prozac was approved in combination with addictive benzodiazepines such as Ativan, Xanax and Valium, but neither the FDA nor Lilly revealed this information. With these patients included, statistical manipulations enabled the FDA to find the drug marginally approvable."

The FDA approved Prozac for depression in December 1987. By 1998 40,000 adverse events including 2,100 suicides had been reported. The most ever recorded in FDA history. **The first Black Box Warning for suicide was not required until Prozac had been on the U.S. market for 17 years.**

Women Beware

Even with the record number of adverse events including thousands of deaths, in November 1999 the FDA's Psychopharmacologic Advisory Committee, with its financial ties to drug companies, recommended Prozac be approved to treat a mental disorder many professionals warn does not exist. Eli Lilly added purple coloring to Prozac (fluoxetine) and called the purple pills Sarafem. The color and name were meant to appeal to women. Lilly then went about crusading for the not-really-new drug Sarafem to be approved for Premenstrual Dysphoric Disorder (PMDD). Sarafem is Prozac, nothing was changed but the color and the name.

Dr. Paula Caplan of Brown University warns PMDD is completely invented by the APA and pharmaceutical industry. She emphasizes giving a mental health label to the natural monthly process will prevent diagnosis of any true health reasons for extreme distress suffered by some women. The level of female hormones estrogen and progesterone vary greatly through the 28-day cycle. That variation can cause some women to have moderate to severe symptoms.

The menstrual cycle is not a mental health disorder. Other countries do not accept the existence of PMDD. Once again the U.S. public is like a herd of cattle to be roped in and branded by the APA and drug industry. The risk that women taking Sarafem may commit suicide apparently is not a concern. No concern that children may be left motherless. Worse yet, mothers on Prozac (or the purple version called Sarafem) have killed their children before committing suicide.

Fraud, Fraud, and More Fraud

New York State Attorney General Eliot Spitzer sued the British-based drug company GlaxoSmithKline (GSK) on June 2, 2004 for persistent fraud. Gardiner Harris of *The New York Times* quoted Mr. Spitzer, "The point of the lawsuit is to ensure that there is complete information to doctors for making decisions in prescribing. The record with Paxil, we believe, is a powerful one that shows that GSK was making selective disclosures and was not giving doctors the entirety of the evidence."

GSK had sponsored five clinical trials with their antidepressant Paxil, testing it on adolescents. GSK did the trials to qualify for a six-month extension of Paxil's patent. The extension is granted under **a U.S. federal law that encourages drug testing on children**. Only one trial was published and it did not show a strong case of the drug benefiting youths taking it. **The four unpublished trials failed to show any benefit and suggested an increased risk of suicide.** An internal GSK memo said the company should have "effectively managed the dissemination of these data in order to minimize any potential negative commercial impact." Physicians are never told about the unpublished trials with extremely important adverse event information.

The widow of Victor Motus lost her lawsuit against Pfizer after Pfizer's attorney Malcolm Wheeler called FDA chief counsel Daniel E. Troy requesting that the FDA intervene in a lawsuit filed against the company for a suicide from the antidepressant Zoloft. Troy had been a Pfizer lawyer just before he was appointed FDA's chief legal counsel. Troy had always been a champion of drug companies.

On November 12, 1998 Victor Motus was scheduled to fly to Washington, D.C. to receive an award from President Clinton for his work in a local school district. Instead Mr. Motus killed himself. Six days before his death he had been given free samples of Zoloft by his family physician. Victor told his wife several times in those six days that the Zoloft was making him "crazy" as he became highly agitated and confused. Mrs. Motus found her husband's body hanging from an attic beam.

In response to Wheeler's plea for FDA help with Mrs. Motus' lawsuit against Pfizer, Troy completely ignored all the scientific findings and filed a legal brief stating that the FDA had dismissed the idea that SSRI antidepressants increase the risk of suicide. The FDA intervention on behalf of Pfizer helped the drug company to not be held responsible. Mrs. Motus received no compensation for the death of her husband.

Internal Pfizer documents show that more than ten years before Victor Motus committed suicide, the drug company was well aware of the risks of Zoloft. In a small trial of the drug during development in the early 1980s, Pfizer tested the drug on healthy volunteers. All the volunteers taking Zoloft dropped out of the trial within the first week because of severe anxiety or agitation.

There are no U.S. government regulations requiring all clinical data must be made public. Instead, the public must suffer death and injury from dangerous drugs, sometimes for decades, and pay billions of dollars from their pockets and in their taxes. Taxes that pay salaries of FDA leaders who protect drug companies.

Decades of Homicides with Links to Antidepressants

"Nervousness, anxiety, self-mutilation and manic behavior" are among the "usual adverse effects" of Prozac according to Eli Lilly documents. The drug company continues to publicly deny Prozac causes violence.

Akathisia is a psychomotor restlessness the FDA identified as an adverse event of antidepressant drugs. The *American Psychiatric Association's Diagnostic and Statistical Manual of Mental Disorders* documents, "**akathisia can drive people to suicide, aggression and violence**"

Dr. Peter Breggin has been working for decades trying to enlighten the government, the medical community, and the public about the dangers of psychiatric drugs. He describes akathisia as "a terrible inner sensation of agitation accompanied by a compulsion to move about. Akathisia is documented as driving people to suicide and to aggression and violence." According to Dr. Breggin, mania, another identified side effect of antidepressants, can cause unusual thought patterns, hallucinations and delusions. The combination of a person suffering akathisia and mania easily leads to senseless violence.

These extreme side effects can also occur as withdrawal symptoms when people stop taking antidepressants. Dr. John Zajecka reported in the *Journal of Clinical Psychiatry* the agitation and irritability experienced by people withdrawing from SSRI/SNRI drugs can cause "aggressiveness and suicidal impulsivity."

"I have thoughts of harming my own children." One man admitted to British psychiatrist Dr. M. Bloch, his extreme side effects while withdrawing from an antidepressant. Dr. Bloch reported in the British medical journal *Lancet* of patients who became homicidal and suicidal after they stopped taking antidepressants.

They are not horrible people. They are people who trusted their physicians to give them safe medications. These are dangerous mind-altering drugs.

Ann Blake Tracey, Ph.D. is a foremost expert of SSRI and SNRI antidepressants. She has been researching for almost 20 years the links between violent crimes and suicide with antidepressants. In her book, *"Prozac: Panacea or Pandora? Our Serotonin Nightmare,"* she describes the dangers of stopping an antidepressant abruptly because the drugs are highly addictive.

There are website forums where people describe how they effectively stopped taking these addictive drugs. Two sites are: http://www.dangerousmedicine.com and http://www.mindfreedom.org/kb/psychiatric-drugs/quitting-psychiatric-drugs.

More than 1,500 incidents of violent crimes have been associated with antidepressants according to the International Coalition for Drug Awareness (ICDA). Between 1997 and 2007 in the U.S. there have been 8 school shootings killing 29 people and wounding 68, when it has been verified the killer was taking antidepressant drugs. This does not include the most recent horrific tragedy at Virginia Tech. Not all the shooters were youths.

In other school shootings the toxicology reports and medical records were not made public so it is unknown if antidepressants were involved, but it is highly likely. This is the first question that should come to your mind when a tragic killing is reported. Was the person taking or had recently stopped taking antidepressants? If that information was publicized in news accounts of every tragic occasion, the public would become aware that antidepressant drugs should be avoided by every man, woman and child. The public needs to stop feeding the pharmaceutical industry's coffers with their dollars and lives.

You do not have to wait for the FDA to take dangerous drugs off the market. You can say no for yourself and your children. As I wrote this chapter, I came across a news account on the Internet about a woman in Texas who killed her three daughters (miraculously a fourth survived) and then herself. No mention was made if she was taking antidepressants. She was reported to be going through a divorce so it is highly probable a physician had given her a prescription for an antidepressant; a prescription for death.

Physicians in all medical specialties hand out prescriptions and free samples of psychiatric drugs as though they are sugar pills. Most physicians are not aware of the extreme dangers of these drugs. As drug companies' internal documents indicate, the drug salespeople aren't telling the physicians.

The dangerous effects of antidepressants have no age limit. The ICDA tracking of violence indicates the age of the psychiatric-drugged attackers ranged from 10 to 61 years old. Examples of well-publicized cases include: Actor Phil Hartman's wife killed him then committed suicide. She was taking Zoloft.

Mark Barton in Atlanta killed his wife, daughter, and son (ages 7 and 11) by bludgeoning them with a hammer and holding them under water. Then he shot to death 9 people in two Atlanta offices, wounding more than 10 others before committing suicide. He was taking Prozac.

Matthew Beck shot 4 co-workers to death then committed suicide. He was taking Luvox.

Jeff Weise shot and killed 9 people and wounded 5 others before committing suicide. He was taking Prozac.

Kip Kinkel killed his parents then went to his school and shot to death 2 students and wounded 22 more. He was taking Prozac.

Kenneth Seguin drugged his two children, ages 5 and 7. He took them to a pond, slashed their wrists and dumped their little bodies in the water. Then he drove home and killed his wife with an ax as she slept. He was taking Prozac.

Barbara Mortenson, 61 years old, was on Prozac for only 2 weeks when she cannibalized her 87 year-old mother.

Patrick Purdy went to an elementary school playground with an assault rifle. His shooting killed 2 children and wounded 22 others. He was taking Elavil.

Adrea Yates drowned her 5 children. She was taking four psychiatric drugs including Effexor.

Samantha Hirt burned her two young children to death. She was taking Effexor.

The New York subway bombing by Edward Leary injured 48 people. He was taking Prozac.

Ten year-old Tommy Becton grabbed his three year-old niece holding her as a shield and aimed a shotgun at a sheriff's deputy. Tommy was taking Prozac.

Steve Lieth shot and killed a school superintendent and wounded 2 others. He was taking Prozac.

Columbine, Colorado teenager Eric Harris was taking antidepressant Luvox when he and Dylan Klebold killed 12 classmates and a teacher and wounded 23 others before committing suicide. The coroner confirmed Luvox was in Harris' system. Luvox is a drug known to cause mania at a high rate in young people. Mania can cause unusual thought patterns, hallucinations and delusions. The toxicology report from Klebold's autopsy was never made public.

The *New York Times* reported investigators of the 2007 Virginia Tech killing spree of Cho Seung-Hui found antidepressant drugs among Seung-Hui's belongings. The toxicology report has not been released. Seung-Hui killed 32 students and faculty members and then committed suicide. Dr. Breggin explains in his book *The Antidepressant Fact Book*, "stopping antidepressants can be as dangerous as starting them since they can cause very disturbing and painful withdrawal reactions."

Don't jump to the conclusion these people were all crazy and that's why they killed. Depression is a natural emotion people have lived with through the ages. Depression doesn't make people violent. Physicians of all specialties are handing out prescriptions for antidepressants as a fix for hundreds of symptoms. **The violence epidemic in the U.S. started with the common use of SSRI/SNRI antidepressants.** People often go through times of depression due to job lose, relocation, loss of a loved one, divorce, and many other situations which cause us to feel insecure. Our bodies do have natural ways of dealing with these emotions especially if people use healthy means including adequate sleep, exercise, healthy eating and emotional support from friends and family.

SSRI/SNRI drugs interfere with the body-mind's normal functioning. These drugs are literally mind-altering. They can cause people to terminate loving, supportive relationships with family and friends, the very relationships that are extremely important to helping people recover from depression. The drugs can cause hallucinations, paranoia, and mania.

There is a direct correlation with the increase of antidepressant drug use and the rise in extreme, senseless violent acts. There are experts who have been trying to bring this to the attention of physicians, the FDA, and the public for more than a decade. **Depression is not the problem. The drugs are the problem.**

In 1998 GlaxoSmithKline was ordered to pay $6.4 million to the surviving family members after 60 year old Donald Schnell flew into a rage and killed his wife, daughter, and granddaughter only 48 hours after he began to take Paxil.

"I keep asking, when is somebody going to see this? But we've been so brainwashed about drugs. We think legal means safe." Dr. Tracey explains. "Most people don't know that LSD once was legal and prescribed as a wonder drug. That PCP was considered to have a large margin of safety in humans. Or that Ecstasy was legally prescribed and sold for five years to treat depression."

The adverse effects of psychiatric drugs are regularly misdiagnosed as more signs of depression, anxiety or some other created-by-vote psychiatric disorder. Then patients are prescribed additional psychiatric drugs or the dosage is increased. That was the case of California teenager Dominique Slater. Only 14 years old, she was on several antidepressants, including Celexa and Wellbutrin. When her erratic behavior worsened, her doctor prescribed double doses of Effexor. Fifteen days later she killed herself. She was barely a teenager yet she was prescribed multiple antidepressant drugs at high doses. The year was 2003. Britain had already sent letters to all physicians sternly warning against the use of any of these drugs in anyone under the age of 18 years. It took the FDA another year to issue a warning of increased suicide in youths under 18 years old. No letters were sent to physicians. And the drug companies created marketing campaigns specifically to get antidepressants into the offices of all types of physicians, not just psychiatrists.

More than 10 million prescriptions for antidepressants are issued each year for children younger than 18 in the U.S. Any physician, not just psychiatrists, can write prescriptions for psychiatric drugs. The age of children being given these powerful mind-altering drugs continues to get younger.

Physicians in Ohio in the month of July 2004 prescribed psychiatric drugs for 696 babies aged newborn to 3 years old covered by Medicaid.

"It's shocking," said Dr. Ellen Bassuk, associate professor of psychiatry at Harvard Medical School. "These medications are not benign. They can have dangerous side effects. Who is being helped by children being drugged, the babies or the caregivers?"

Scientific Evidence of Antidepressants' Effects on Newborns

"When we put pregnant women on antidepressants, they can't get off them." An unconcerned gynecologist told my friend C. when she told him she had spent years trying to get off the antidepressant he had prescribed to her. C. suffered suicide ideation while on the drug. Three years before this callous physician's comments to C., the extreme health risks to the fetus had been reported in medical journals.

A study published in *The New England Journal of Medicine* in February 2006 reports pregnant women taking antidepressants have babies who are 6 times more likely to have primary pulmonary hypertension (PPH) or a developing lung disorder. PPH is extremely serious. A baby's organs such as brain, kidney and liver are stressed due to lack of oxygen. PPH requires neonatal intensive care. PPH can be fatal, and for babies who do survive there is often long-term health problems including breathing difficulties, seizures and developmental disorders.

Women taking SSRI/SNRI drugs during the first trimester of pregnancy are at 60 percent greater risk of their babies having heart defects and 40 percent greater risk of their babies suffering malformation.

"In conclusion, our results suggest that maternal exposure to fluoxetine (Prozac, Luvox, Sarafem and Symbyax) during pregnancy and lactation results in **enduring behavioral alterations...throughout life**." A study reports in *Pharmacology*, spring 2007. There is nothing preventing drugs a pregnant woman takes from going directly into the bloodstream and then all the tissues of the fetus. And as this study indicates, antidepressants are also transferred to the baby through the mother's milk.

As of July 2009, the drug companies, using their puppets and financial influence, were heavily lobbying the U.S. Senate to pass a bill called the Mothers Act. This insane bill has already passed the U.S. House of Representatives. Supposedly the Act is meant to address post-partum depression. The truth is that it is the drug industry influencing legislation in order to have more taxpayers' money flow into drug companies' profits. The 1,100 drug industry lobbyists on Capitol Hill are greasing the skids well so that this dangerous legislation that will harm, not help, mothers, babies, and American families will easily pass. It's about money, not health.

Senator Robert Menendez of New Jersey is the main sponsor of the bill in the Senate. According to the public interest group, Common Cause, Senator Menendez received over $2 million from the healthcare industry, including drug companies, from 2000 to 2008. New Jersey now has a law forcing all new mothers to submit to mandatory mental health screening; forced screening for disorders that are voted into existence! The health risks, including a high rate of death, to the mothers, babies, and families are of no concern to those in government who are financially tied to the high-paying pharmaceutical industry.

Children and Teens on Medicaid Are Especially Susceptible

Drug salespeople are trained and expected to encourage physicians to write prescriptions when public money is paying for the drugs. I know this from my own experience. The company I worked for had a marketing campaign specifically aimed at Medicaid. Those opportunities for prescriptions were called the "low hanging fruit." We were instructed to always remind physicians to prescribe our drugs for their Medicaid patients. It was less hassle for the doctors. They would not have to justify the prescriptions to private insurance companies. And when people do not have to pay for their medications, they usually continue to take them. That means continued sales for pharmaceutical companies. Medicaid serves as a cash cow to the drug industry.

In the case of all the psychiatric drugs, children and teens on government paid programs are especially susceptible. A review of Medicaid records discovered that **in Texas in just the** months of July and August 2004, 63,118 teens were prescribed psychiatric drugs and billed to a public funded program. The review also found one-third of the children were on multiple mind-altering drugs. Many of the drugs are not approved for children or adolescents. Even the drugs that are approved, like Prozac, have consistently been shown as unsafe and ineffective.

Ohio Medicaid spent more than $65 million on psychiatric drugs just for children in 2004. Nearly one-third of the children ages 6 to 18 in foster and group homes were being given psychiatric drugs. Some of the children were on five or more of these powerful mind-altering drugs.

In Tennessee children covered by TennCare were put on non-approved antipsychotic drugs. The number of children given these drugs doubled in six years, including a 61 percent increase in prescriptions for preschool children.

The *Boston Globe* reported in August 2004 two-thirds of the children in state care in Massachusetts were taking antipsychotic drugs. Many of the drugs are not approved for children. All are addictive and have extreme side effects.

The Florida Department of Children and Family Services analyzed records for the year September 2002 to September 2003 and found 41,993 children ages 12 and younger were given 190,210 off-label prescriptions for psychiatric drugs by physicians in all fields of medicine including geriatrics, anesthesiology, radiology, dermatology, ophthalmology, obstetrics, pathology and other areas of medicine.

"We are taking away their future." Dr. Tony Appel, a neuropsychologist (brain specialist) and expert child advocate, told NBC News. "We are taking away their ability to relate to people. Trust, love, caring, the ability to put yourself in the other person's shoes and to see how they see you, we take all that away from these children."

Why do doctors continue to believe drug companies' sales pitches? Do we really need door-to-door selling of drugs? While children and adults are dying from psychiatric drugs, pharmaceutical representatives and managers are rewarded with bonuses. That is true insanity.

Have Americans become so obsessed with instant gratification that they choose to drug children rather than love them and look for root causes of problems? Why do we allow the drug industry and the psychiatric professionals with whom they have financial ties to cause so much damage and pain while they make billions of dollars annually from these dangerous drugs?

The current score in this life and death game:

Score: Poor Children – Zero, Drug Companies – Won

Score: Tax Payers – Zero, Drug Companies – Won

16

THE SCAM OF MENTAL HEALTH TESTING IN SCHOOLS

Mental health testing of school children without parental consent is here in the U.S. now. The New Freedom Commission (NFC) is a group created by a presidential executive order in 2002. The NFC's members are from the drug industry or have financial ties with it. They designed an action plan for "early mental health screening, assessment and referral to services." The NFC recommends mandatory mental health screening for all high school students. As history has shown, once a federal program is created it becomes institutionalized and rarely goes away. The costs are rushing downhill towards the public like a thousand Niagara Falls; costs in lives and money.

The American Association of Physicians and Surgeons (AAPS) has called the New Freedom Commission's plan "a dangerous scheme that will heap even more coercive pressure on parents to medicate children with the potential of dangerous side effects."

The New Freedom Commission (NFC) encourages mental illness screening of all children and adolescents under the guise of preventing youth suicides. The Commission completely ignores the U.S. Preventive Service Task Force finding that there is no evidence screening for suicide risk reduces suicide attempts or mortality.

The National Institute of Mental Health (NIMH) Suicide Research Consortium study found, "...**a prevention program designed for high-school aged youth found that participants**

were more likely to consider suicide a solution to a problem after the program than prior to the program."

Jane Pearson, Ph.D. who chaired the NIMH consortium stated that when researchers have tried to predict suicide using as many known risk factors as possible (not like the simple TeenScreen test) they are still unable to predict who will and who will not attempt suicide.

TeenScreen Finds Most Teens Are Mentally Ill

The Federal government has budgeted money to bribe states to follow NFC guidelines. The screening process the NFC chose as the model, TeenScreen, identified 50 percent of teenagers who were screened in Colorado high schools as suicidal and having mental disorders. The fact that **9 out of 10 children or adolescents who see psychiatrists are put on drugs** should be reason enough to have all Americans shouting, "NO MORE DRUGGING OUR CHILDREN!"

Say no to physicians giving your children psychiatric drugs. Say no to schools that try to coerce you into drugging your child. Just say no to drug companies. Stop believing their puppets. Children need love and healthy environments, not psychiatric drugs. The same is true of adults. Remove people's feelings of helplessness and isolation and watch depression disappear, if it was ever actually there.

Parental Consent Is Not Needed

One of the scare tactics the government and pharmaceutical companies use to promote screening teenagers for mental illness is that suicide is the third leading cause of death in teenagers. The fact is, for the most part teenagers are healthy because they are so young. They do not have aged bodies with impaired hearts and organs. Accidents are the leading cause of death of adolescents.

When you look at the real numbers, suicides of children and teens are rare. About 10 in 100,000 teens ages 15 to 19 commit suicide when they aren't on antidepressants or other psychiatric drugs, including ADD/ADHD drugs. Instead of reducing suicide, **psychiatric drugs increase the suicide rate**. This is clearly shown in a study published in the *Archives of General Psychiatry*, August 2006. It found that children 6 to 18 years old in an inpatient setting who were taking antidepressants were 52 percent more likely to attempt suicide within 2 months of discharge than children not taking the drugs. Why are any children given these drugs? Since 2004 there has been a Black Box Warning for increased risk of suicide for under age 18. And then in May 2007 the FDA increased the age to younger than 25 for the Black Box Warning. It applies to all SSRI/SNRI antidepressants.

This is a good example of government, with financial encouragement from the drug industry, creating a new system because of special cause variation. Teenage suicide is tragic, but it is such a low number compared to the total number of teens that it is special cause variation. New systems to address special cause variation are an unnecessary expense. A system to screen every teenager in the country because 10 in 100,000 teens commit suicide is like you buying an expensive race car to get to work because an accident one day caused an exceptionally long commute time. And besides, this is just the façade. The truth behind the government's actions is to give drug companies access to children, just as the Texas state government did. **Drug companies and the APA are preying on American children. It is not about health. It is about money.**

Besides the billions of dollars this initiative is costing taxpayers, this screening process will also create immeasurable costs. Labeling teens mentally ill or suicidal can follow them throughout life. This important fact was included in a bill put to the U.S. House of Representatives in 2005, 2007, and again in 2009 by Representative Ron Paul, a physician. "Federal funds should never be used to support programs that could lead to the increased over-medication of children, the stigmatization of children and adults as mentally disturbed based on their political or other beliefs or the violation of the liberty and privacy of Americans by subjecting them to invasive 'mental health screening' (the results of which are placed in medical records which are available to government officials and special interests without the patient's consent). "

Federal funds are being spent. In 2005 approximately $18.8 million was made available to states in the form of Mental Health Transformation Grants. In 2006 that amount almost doubled to $36 million. The grant originated from the New Freedom Commission (NFC). The Commission created the Garrett Lee Smith Act, signed into law October 21, 2004. The Act gives grant money to implement teen and youth mental health screening and other recommendations of the NFC. The Substance Abuse and Mental Health Services Administration (SAMHSA), an agency within the U.S. Department of Health and Human Services, is the "lead government agency overseeing the goals and recommendations of the NFC" according to their website. SAMHSA awards the Mental Health Transformation Grant money to encourage schools to establish mental health screening.

Suicide is a constant theme throughout the documents of SAMHSA and the NFC. As I have discussed, the suicide rate for teens not taking antidepressants is low. Suicide rates increase with age. Just how many Americans a year actually commit suicide? About 31,000 people of all ages in the U.S. commit suicide every year. Yes, there are others who try but do not succeed. But death by suicide that the NFC and its drug industry supporters are using to create fear in Americans is not at a magnitude that all the promoting of screening justifies. In contrast, **more than 106,000 Americans die every year from prescription drug side affects.** And that is the reported number. The actual may be twice the official. The suicides and homicides due to antidepressant drugs usually are not included in the official numbers.

As a taxpayer, how do you feel about that? Your tax dollars are being spent to bribe states to allow ineffective mental health screening in schools; a screening method so invalid that it identifies more than half of teenagers as having created-by-vote mental health disorders. In the next chapter I provide the history of what happens when schools are bribed with federal money. The drug companies have done it once. There is nothing to prevent another mass drugging of our children unless America wakes up and shouts no!

Pharmaceutical's Death Grip

TeenScreen was developed at Columbia University with drug industry funding. **"It's just a way to put more people on prescription drugs,"** states Dr. Marcia Angell of Harvard Medical School and author of *The Truth About Drug Companies.* Dr. Angell is the former editor of the *New England Journal of Medicine.*

As of June 2009, TeenScreen had 542 active screening sites in 43 states. The New York public relations firm Rabin Strategic Partners was hired by TeenScreen to enhance its image. New materials emphasizing the word "science" (which there is none of with TeenScreen) were designed and a public relations campaign is in progress. The drug industry is dressing-up one of their puppets.

The TeenScreen Newsletter instructs schools to make the TeenScreen survey a part of the curriculum to ensure that parental consent is not required before children are screened,

"...if the screening will be given to all students, as opposed to some, it becomes part of the curriculum and no longer requires active parental consent." Parents, how do you feel about that? Do you want the drug industry preying on your children in schools?

Even if the testing is not part of the curriculum, screenings of teenagers is done without explicit parental consent. TeenScreen depends on passive consent, a method where a consent form is sent home with the child. If the form is not returned stating parental objection it is assumed the parents agree to the mental illness screening of their child. TeenScreen entices youths to take the test by offering coupons for free videos and pizza.

Representative Ron Paul's bill H.R. 2218, also called *The Parental Consent Act of 2009*, forbids federal funds from being used for any mental health screening of students without the express, written, voluntary, and informed consent of parents. The bill states: "The United States Preventive Services Task Force (USPSTF) issued findings and recommendation against screening for suicide that corroborate those of the Canadian Preventive Services Task Force. USPSTF found no evidence that screening for suicide risk reduces suicide attempts or mortality."

The bill also presents the Surgeon General's report: "The 1999 Surgeon General's report on mental health admitted the serious conflicts in the medical literature regarding the definitions of mental health and mental illness when it said, 'In other words, what it means to be mentally healthy is subject to many different interpretations that are rooted in value judgments that may vary across cultures...there is no definitive laboratory test or abnormality in brain tissue that can identify the illness'." There is no science to anything labeled a mental disorder. Mental disorders are figments of APA and drug industry imagination.

The bill makes the important point, "A September 2004 FDA hearing found that more than two-thirds of studies of antidepressants given to depressed children showed that they were no more effective than placebo, or sugar pills, and that only the positive trials were published by the pharmaceutical industry."

The bill provides significant facts about antidepressants:

"The FDA finally acknowledged in September 2004, that the newer antidepressants are related to suicidal thoughts and actions in children and that this data was hidden for years. The FDA had over two thousand reports of completed suicides from 1987 to 1995 for the drug Prozac alone, which by the agency's own calculations represent but a fraction of the suicides. Prozac is the only such drug approved by the FDA for use in children."

This important bill that would give the American public some protection from the drug industry cartel was referred to committees. The process in Congress is after a bill is introduced it is sent to committees for further analysis. Most bills never see the light of day after that. In the case of this important Parental Consent Act, never getting on the floor for debate and recorded vote is an easy out for the members of Congress. They won't anger the pharmaceutical companies that gave millions of dollars to their campaigns, or their constituents who would expect them to vote for a sensible bill that upholds the rights of parents trying to protect their children, and decreases the amount of tax dollars flowing into pharmaceutical coffers.

The entire bill is provided in Appendix B. It will only see the light of day if enough citizens notify their representatives that they want H.R. 2218 supported and voted on. Then it will still need to be passed by the Senate and signed into law by the President. All this expense for something that should be an inalienable right in the U.S.; parents making health decisions for their children.

When Representative Ron Paul presented the bill to the House in 2009, he told his fellow Representatives:

> "Already, too many children are suffering from being prescribed psychotropic drugs for nothing more than children's typical rambunctious behavior. . . Many children have suffered harmful side effects from psychotropic drugs. Some of the possible side effects include mania, violence, dependence, and weight gain. Yet, parents are already being threatened with child abuse charges if they resist efforts to drug their children. Imagine how much easier it will be to drug children against their parents' wishes if a federally-funded mental-health screener makes the recommendation."

If you are a parent or guardian of children who are in a public school system, talk with them about the mental health screening. Let your children know what you want them to do. Warn them about the bribing with free gifts TeenScreen uses to entice children to take the screening. Then you won't be the last to know your child has been labeled with a mental health disorder. Talk with your school board. Find out if TeenScreen has already infiltrated your child's school. Join with other concerned parents and anyone in the community who recognize the grave danger. This is the pharmaceutical industry preying on the children of this country. As a community you can prevent TeenScreen from being accepted by the school or have TeenScreen removed if it is present in the school. Taking action at a grassroots level is within every parent's power and is needed throughout the country. We dare not wait for the Parental Consent Act to pass.

The parents of 16 year-old Chelsea Rhoades of Indiana had no idea her school had allowed TeenScreen to test in the school. Chelsea was given a ten- minute TeenScreen mental health test in her high school, and based on it was told she had two mental health disorders. Chelsea's parents took quick action. They filed lawsuits against the school officials who allowed the test to be administered without parental consent and the TeenScreen program.

Physicians can also do their part to stop the preying of drug companies on children. As Voltaire so eloquently advised, "If you can get people to believe in absurdities, you can get them to commit atrocities." The drug companies wine and dine physicians to convince them to believe absurdities. In Nevada on June 30, 2009, Senator Harry Reid served as the host for a dinner for invited physicians. A brochure distributed at the dinner sold the absurdity, "TeenScreen Primary Care – Making Mental Health Checkups A Primary Care." The presentations made at the dinner by TeenScreen promoters were to convince primary care physicians and pediatricians to include TeenScreen in their practices. One paid speaker described how TeenScreen evaluations made by physicians could be billed to insurance companies. Another speaker, a physician, boasted it took him only one minute to do a mental health screening. One minute to label a child with a contrived mental disorder (absurd) and then put them on drugs (atrocious). Unfortunately a majority of the physicians attending the dinner appeared to accept the absurdity and agreed to incorporate TeenScreen into their practices.

If you live in the United States of America your children can be given ineffective screening without your consent for what the professionals admit is a subjective condition called "mental health." This is literally the death grip that the pharmaceutical and psychiatric industries have on the U.S. But while drug companies have significant control of the political and medical establishments, you do not have to allow them to control your life or your children's lives. Don't let anyone turn you or your child into a prescription drug addict or a corpse with psychiatric drugs.

Drug companies are spending hundreds of millions of dollars to ensure the New Freedom Commission guidelines they wrote are fully implemented. If we project the Colorado numbers to the national level we see why. In the U.S. there are 19,800,000 youths between ages 11 and 18. If all youths were screened, applying the 50 percent rate from the Colorado schools, 9,900,000 youths would be labeled as having psychiatric disorders. And since 9 out of 10 youths who receive psychiatric treatment are put on drugs, that would mean 8,910,000 youths would very likely be given highly addictive drugs that cause suicide and violence. Pharmaceutical companies would make billions of dollars every year from drugging children and creating potential life-long drug users, until the children commit suicide.

Aliah's Story

In Texas, where this all began, some teenagers were put on drugs without the consent of their parents. Social services removed some teenagers, like 13 year-old Aliah Gleason, from their homes without cause other than their parents did not agree to put their children on psychiatric drugs. Aliah had not committed any crimes. She is a vocal teenager who had "mouthed off" to some of her teachers. Medical records show Aliah was given 12 different psychiatric drugs during her incarceration in a state mental health facility where she was held without her parents' consent. Her parents were not allowed to visit her during five of her nine months of involuntary incarceration. According to her medical records her diagnosis was never clarified and changed multiple times as the drugs she was forced to take were changed.

Drugs are not the answer. They are the problem. Dr. John Breeding has helped Aliah safely get off all the psychiatric drugs and she has returned to living a normal teenager's life.

You are not powerless to the pharmaceutical behemoth. Find out if TeenScreen has invaded your community's schools. Let your voice be heard for the sake of your children or the children in your community. Check the Internet for organizations that can assist your grass-roots efforts to keep TeenScreen out or to remove it if it is already being used in the schools.

If you truly have a heart, check if there are children in homeless shelters in your city. Children in shelters and orphanages are often targeted by the predatory drug industry and their partner TeenScreen. At a Colorado youth homeless shelter, TeenScreen labeled 71 percent of the youths with psychiatric disorders. Who would have a happy state of mind living in a homeless shelter? Homeless children don't need to be labeled with psychiatric disorders. They need homes and a sense of security. No drugs will provide those. The labels given these children will follow them for life. And the psychiatric drugs will take their health, and maybe their lives.

Medicaid will pay for the drugs ensuring that prescriptions for these highly addictive drugs are handed out to homeless youths, as the evidence from several states has shown. Physicians are prescribing highly addictive, dangerous mind-altering drugs for symptoms of youths feeling unwanted and unloved. What kind of society does that?

In an authentically strong nation every child matters. Children are the future of every society. We must take steps to stop the drug industry creating a society of drug addicts. The U.S. is well on its way to having a large percentage of the primary working population, ages 20 to 55, using psychiatric drugs. That will keep the drug companies rich, but it will make a very poor, weak nation.

Appeal to physicians in your community. They hold the power of the pen to write or refuse to write prescriptions for these drugs. Communities need to open communication with the physicians that practice in their area. Every working day drug representatives visit physicians, plying them with gifts and using selling skills to convince doctors to write more prescriptions. Communities can find common ground with physicians. Most doctors don't want to give children drugs that cause addiction and possible death. The common ground should be the shared aim of healthy children.

"TeenScreen is nothing more than a government-sponsored marketing tool created to serve the interests of the corporate pharmaceutical industry and psychiatrists. It is a shame and disgrace that the United States is putting millions of children on psychiatric drugs today." Is the emphatic message of Dr. John Breeding, psychologist, author and director of Texans for Safe Education.

Score: Parents – Zero, Drug Companies – Won

The recommendations of the NFC completely ignore that the voted-in symptoms of mental disorders are symptoms of measurable and testable physical ailments. TeenScreen completely ignores this and has nothing in its process to suggest that youths first see their family physician for physical health evaluations. Of course the TeenScreen survey is so invalid that a perfectly healthy teenager has a 50 percent chance or greater of being labeled mentally ill.

The Frightening Truth

Once a person is taking a psychiatric drug s/he will have a difficult time stopping. Drs. Baughman, Breggin, Breeding, Tracey and many more respected experts have tried to warn the public and physicians that these drugs are extremely addictive. The drug companies know it. They bank on it because addiction means continued sales. The more addicts created while they are children and teenagers, the more adults who will be continually taking drugs. If not the psychiatric drugs, then drugs for the diabetes some psychiatric drugs cause. Or drugs for the heart disease some psychiatric drugs cause. Or drugs for any number of serious health problems that are side effects of psychiatric drugs – if the person doesn't commit suicide. And suicides leave grieving family members who may very likely be given prescriptions for antidepressants. That is the frightening truth the drug companies do not want the American public or any other country to realize. That is why the drug companies approached Texas about mental health. It is impossible to define because there is no scientific basis for mental illness and the drugs are all highly addictive. The drug companies can't lose. They are creating birth-to-grave use of drugs.

Drug companies are some of the wealthiest companies in the world. They have used that wealth for self-interest gain. They have purchased influence at all levels of medical care and at all levels of government. They are allowed to be an **industry that has deceived the public with intent to harm**. The drug industry has been getting everything it wants from the federal and state governments in the U.S. The pharmaceutical industry wrote Medicare Part D the way they wanted it. The CBS Sixty Minutes report "Under the Influence" showed the strong arm-twisting by drug lobbyists to force members of Congress to pass the bill. The public is now strapped with an excessively expensive program that is costing many elderly more for their medicines.

The drug industry wants American children drugged. The government is helping them do it. That is why **the pharmaceutical industry is the greatest threat to U.S. healthcare.**

"You see a fairly significant percentage of patients where new and more severe psychiatric symptoms are triggered by the drug itself...It's brilliant from the capitalist point of view. You take a kid and you turn them into a customer, and hopefully a lifelong customer." Robert Whitaker, M.D. explains in "Anatomy of an Epidemic: Psychiatric Drugs and the Astonishing Rise of Mental Illness in America." The article was published in *Ethical Human Psychology and Psychiatry Journal*.

Score: The People – Zero, Drug Companies - Won

The Federal Government Quietly Takes More Control

The U.S. Department of Health and Human Services Subcommittee on Children and Family published policy options on February 5, 2003 not long after the creation of the New Freedom Commission. The policy is titled "Promoting, Preserving, and Restoring Children's Mental Health." A few of the many implementation options the subcommittee recommends are:

- Strengthen mental health services in schools and school's role in promoting social and emotional well-being
- Create a state level infrastructure for school based mental health services
- Train teachers and school personnel to recognize signs of emotional problems in children and to make appropriate referrals for assessment and services

America, please wake up and smell the baloney. This may sound as though it is for the good of children, but it is merely the drug industry expanding its lucrative gold mines to the national level. When screening in schools was initiated in Texas, prescriptions for psychiatric drugs jumped. And the Texas Mental Health and Mental Retardation system went bankrupt.

The aim of schools should be education. Over-worked teachers do not need to be trained to detect emotional problems. The fact that physicians often misdiagnose authentic physical problems as mental health problems is justification enough to not waste resources trying to train teachers to become health diagnosticians. Their job is to be educators. If children realize teachers are watching them for signs of mental problems it will be detrimental to the teacher/student relationship. Healthy relationships of all types grow from people feeling others are doing something *with them*, not *to them*, or *for them*. Very often children are put in positions where things are done to them. It increases their sense of helplessness and does not foster positive emotional health.

"If depression is a product of our conflicts, stressful life experiences, and stifled choices, a drug would have no direct effect on treating it." States Dr. Breggin who has continually spoken out against the government's plans for mental health screening of teens and children.

Dr. Elizabeth Roberts, a child and adolescent psychiatrist for more than 30 years and author of *Should You Medicate Your Child's Mind*, felt so strongly about the changes she saw in her profession that she published an opinion in the *Washington Post*. "The changes I have seen in the practice of child psychiatry are shocking. There was a time when doctors insisted on hours of evaluation of a child before making a diagnosis or prescribing a medication." She said parents sometimes tell her of a five-minute appointment with a pediatrician (not pediatric psychiatrist) that ends with the child being given a prescription for a psychiatric drug. Even of her own profession she says, "Psychiatrists are now misdiagnosing and over-medicating children for ordinary defiance and misbehavior. The temper tantrums of belligerent children are increasingly being characterized as psychiatric illnesses. Using such diagnoses, doctors are justifying the sedation of difficult kids with powerful psychiatric drugs that may have serious, permanent or even lethal side effects."

Barry Tuner, a British professor of law and medical ethics, said drug companies have caused the illusion of mental illness to skyrocket in the U.S. and because of it the U.S. "is facing a societal catastrophe...In twenty years a huge percentage of the population will be damaged by these medications and the recipients will have real mental disorders caused by the drugs."

How have drugs that clearly are not effective and have extremely serious side effects, including death at a phenomenal rate, been approved and then fraudulently marketed for decades in the U.S.?

Lawrence Diller, M.D., pediatrician and author of the book *Should I Medicate My Children,* testified before an FDA advisory committee in September 2004. Regarding the conduct of drug companies concealing adverse effects of drugs and marketing off-label prescribing, Dr. Diller poignantly stated:

"The blame is clear. The money, power and influence of the pharmaceutical industry corrupt all. The pervasive control that the drug companies have over medical research, publications, professional organizations, doctors' practices, Congress, and yes, even agencies like the FDA, is the American equivalent of a drug cartel."

States Are Being Bribed

The New Freedom Commission and the drug industry are bribing states to follow the Texas model. In Pennsylvania, two state investigators have sued the state after they claimed they were fired for exposing the drug industry's extreme influence over state prescribing practices. The whistleblowers are concerned the TMAP and TCMAP inappropriately medicate patients, especially children. Allen Jones is one of the whistleblowers. He was an Investigator in Pennsylvania's Office of Inspector General (OIG), Bureau of Special Investigations. His experience is posted on the Internet. He describes what happened when he tried to "expose evidence of major pharmaceutical company wrongdoing...As I attempted to explore and surface these facts I met stiff resistance by OIG officials. I was told that pharmaceutical companies are major political contributors and that I should not continue my probe. The more I attempted to delve, the more I was oppressed by my supervisors. I was effectively threatened with loss of job, career, and reputation if I continued to investigate the pharmaceutical companies."

We do not need a crystal ball to see what will happen if Americans sit quietly and allow the government to promote mental health screening in schools by offering money to states. **The country has been down this dangerous and deadly road before.**

17

UNITED STATES
LAND OF DRUGGED CHILDREN

July 19, 2007

"I saw the teacher give a boy a double dose of Ritalin," the middle-aged woman told me gravely as we waited for a flight in the Raleigh airport. She explained that she worked as a teacher's assistant in another state. I was on my way to a writers' conference, carrying a preliminary copy of this book. We talked about the horrors of drugging children in the schools. This brave woman had gone to school authorities when she saw a teacher giving children more Ritalin than they were prescribed to receive. "I hope your book will be published. Parents need to know how wrong it is to drug children." She said as we boarded our flight.

Children have always been at the mercy of adults. Throughout the ages children have had to endure unhealthy cultural customs. For centuries in China the feet of baby girls were bound tightly, preventing their feet from growing normally, and as adults the women often could not walk without assistance. At least it was not fatal. We now view this type of cultural action as abusive. One hundred years from now how will people view the drugging of American children? It has been fatal. Drugging children because they are active is unique to the U.S. culture. Other countries do not even recognize Attention Deficit Disorder (ADD) and Attention Deficit Hyperactive Disorder (ADHD).

"We do not have an independent, valid test for ADHD, and there is no data to indicate that ADHD is due to a brain malfunction." The National Institute of Health's consensus statement clearly indicates that ADHD is a contrived label; a contrived label that rakes in billions of dollars annually.

Just the fact that the U.S. is the only country that finds children's energy and inquisitiveness to be unnatural should be a red flag. ADD and ADHD are two more created-by-vote psychiatric disorders the drug industry generously helped to define and promote. The American Academy of Child and Adolescent Psychiatry (AACAP) has recently voted for new ADHD diagnosing guidelines. Two of the principle creators of the guidelines have financial ties to several drug companies that market drugs for ADD/ADHD. Guidelines are changed in order to rope in more unsuspecting parents to drug their children. It's about money not health.

All the drug company puppets are used with ADD/ADHD, including a support group, Children and Adults with Attention Deficit/Hyperactive Disorder (CHADD), that is financially supported by drug companies. But the puppets are not making children laugh. They are creating living childhood nightmares. And death. The nightmares then continue for the grieving families.

Steven Plog is a former CHADD coordinator. He describes in the documentary film "Making a Killing" how he helped a group of children and their parents improve their health with better nutrition and counseling. The children got off the dangerous, addictive pharmaceuticals and everyone was healthier and happier. CHADD and their drug industry supporters were not pleased. Mr. Plog received a letter from CHADD's national office telling him, "We (CHADD) do not allow people to talk about nutrition. We only prescribe people to go to psychiatrists who will then prescribe drugs." The letter told Mr. Plog he was no longer needed as a CHADD coordinator.

Millions of Children Are Being Given Drugs Equivalent to Cocaine

The U.S. Drug Enforcement Agency (DEA) ranks methylphenidate (Ritalin) and amphetamine (Dexedrine and Adderall) in the same Schedule 2 of the Controlled Substances Act as opiates, cocaine and barbiturates. **Schedule 2 includes only drugs with the very highest potential for addiction and abuse.** More brand names of these powerful drugs are: Attenta, Concerta, DextroStat, Desoyxyn, Focalin, Metadate, Methylin and Rubifen. All drugs prescribed for ADD/ADHD are DEA Schedule 2 drugs.

A transdermal patch called Daytrana was approved by the FDA in April 2006. The FDA had rejected the application for Daytrana approval in 2002 due to the high incidence of side effects. In clinical trials 61 percent of children suffered appetite loss and 47 percent suffered insomnia. Two basic needs of human development are wholesome foods and sleep. Drugs do not help children to develop physically and mentally. The drugs impair healthy development!

Novum, maker of the patch, did new trials leaving the patch on 9 hours rather than 12. There was still a high incidence of side effects in the children including nausea, poor appetite, insomnia and physical tics. The FDA chief medical reviewer initially rejected again, but then changed his mind and approval was awarded. The patch is still a Schedule 2 category drug. Would you put cocaine patches on children so they have cocaine continuously getting into their young systems for 9 to 12 hours every day for years?

The U.S. uses more than 80 percent of all the methylphenidate produced every year. And approximately 90 percent of that is prescribed to children. Without ADD/ADHD there is no market for these drugs. Methylphenidate and amphetamine existed first. The "disorders" were needed to create a market. Children as young as 18 months old are being turned into drug addicts with immeasurable damage to their forming systems for no reason other than they are active. And for many adults active children are inconvenient.

Dr. Peter Breggin testified before a U.S. Congressional Committee on Education:

> "Hundreds of animal studies and human clinical trials leave no doubt about how the medication works. First, the drugs suppress all spontaneous behavior…this is manifested in a reduction in the following behaviors:
> 1) exploration
> 2) curiosity
> 3) socializing
> 4) playing"

Dr. Breggin explained to the committee the adverse stimulant effects of the drugs are commonly mistaken by adults as improvement in the child. The children are easy to control when they are drugged and for many adults control is a priority. How would you like to be drugged so you can be controlled?

That is why the drug industry and psychiatry chose to design disorders for children. Children are at the mercy of adult authority. The disorders ADD/ADHD were created in the early 1960s right at the time when the psychedelic agents LSD and mescaline were named by the DEA as dangerous and their sale or possession made illegal. The drugs methylphenidate (Ritalin and others) and amphetamine (Dexedrine, Adderall and others) have a hideous, inhumane history.

History of Methylphenidate and Amphetamine

American psychiatrists had been using LSD and mescaline from 1949 until they were illegalized in the 1960s to perform inhumane research. Instead of testing on lab rats to investigate psychosis, they gave the drugs to humans just to observe the results. Dr. Paul Hoch began the experiments in 1949 with no regard for the recently signed Nuremberg code. He said the drugs "heightened the schizophrenic disorganization of the individual." The drugs were used to increase mental patients' psychotic episodes. When LSD and mescaline became illegal psychiatrists switched to using amphetamine and methylphenidate.

During World War II amphetamine inhalers were given to soldiers to combat fatigue and increase alertness. After decades of reported abuse, the FDA banned the inhalers and limited amphetamines to prescription use in 1965. Psychiatrists were happy to have a new psychiatric drug to replace LSD and mescaline.

Dr. David Janowsky recorded amphetamine injections "rapidly intensify psychotic symptoms." He used several drugs to observe which had the strongest effect on people. Methylphenidate caused a doubling in the severity of symptoms. NIMH records of one man injected with methylphenidate described the terrible, painful reactions to the drug: "Within a few minutes after the methylphenidate infusion, Mr. A. experienced nausea and motor agitation. Soon thereafter he began thrashing about uncontrollably and appeared to be very angry, displaying facial grimacing, grunting and shouting. Pulse and blood pressure were significantly elevated . . . Fifteen minutes after the infusion he shouted, 'It's coming at me again – like getting out of control. It's stronger than I am.' He slammed his fists into the bed and table and implored us not to touch him, warning that he might become assaulting."

Not one of the thousands of people used as human guinea pigs received any benefit from the drugs. Most suffered increased mental health problems because of the drugs. These Nazi-like experiments on humans were allowed to go on in the U.S. from 1949 until 1998. Then a woman Holocaust survivor, Vera Sharav, and Adil Shamoo, Ph.D., a biology professor whose son suffers with schizophrenia, began to make the public aware of this psychiatric "research" that goes against the Hippocratic oath, "first do no harm," and the Nuremberg Code written by two American physicians at the end of World War II, that declares the interests of science should never take priority over the rights of the human subjects and informed consent is always required.

Yet, in the U.S. for 50 years, psychiatrists were allowed to conduct drug experiments with intent to harm and without informed consent on vulnerable people who were seeking professional help. These cruel actions were not conducted in secret. The National Institute of Mental Health funded the research called "symptom-exacerbation experiments." The psychiatric community knew about the research through the published results. The inhumane research was conducted all over the U.S. including elite institutions like Yale University.

"The worse sin toward our fellow creatures is not to hate them, but to be indifferent to them; that's the essence of inhumanity."
George Bernard Shaw, 1897

Human ego is the source of the indifference required to treat another human with such disregard. Nazi physicians experimented on captives in concentration camps because Germans believed they were a supreme race. Other races could therefore be sacrificed for the good of the better race. The U.S.

psychiatrists who performed inhumane research with mind-altering drugs thought of the patients as lesser people than themselves. These physicians had no concept of the workings of the spirit-body-mind, of the whole being.

The drugs were never shown to benefit any person diagnosed with a mental disorder. In fact, the purpose of the experiments was always to increase psychotic episodes, to make the patients suffer more. As Vera Sharav and Adil Shamoo made the media, Congress, and the public aware of the ghastly research it seemed to end in 1998. But the drug companies had already used their financial influence with the government to sell these harmful drugs for use on another vulnerable group in society, children.

American children are continuously given the message taking drugs is a natural aspect of living. That is the message children receive from being told to start the day taking a drug, or watching their friends at school being given drugs by teachers, or seeing their parents pop prescription pills, or watching television and being inundated with drug commercials, or all of the above. The message is clear to children – drugs are as natural and necessary for everyone to take as eating food. But drugs are not natural. What is the message you want your children or grandchildren to learn?

We cannot tell children to "just say no to drugs", meaning illegal drugs, while giving children addictive, mind-altering drugs just because the drugs are labeled legal. LSD was legal. So was Ecstasy. Mind-altering drugs are mind-altering drugs. The death rate for legal drugs in the U.S. is more than 106,000 people a year. And since that number is based on government databases the actual number may be closer to 200,000. The death rate from illegal drugs is a fraction of that, approximately 20,000 a year. You should not trust the safety of a drug just because it has been labeled "legal."

"The DEA has noted serious complications associated with Ritalin (methylphenidate) including suicide, psychotic episodes and violent behavior," Colorado State School Board member Patti Johnson told the school board in 1999. "The NIH also reported that Ritalin and other stimulant drugs result in 'little improvement in academic or social skills,' and they recommend research into alternatives such as change in diet or biofeedback."

Children are naturally active. It is the way they are meant to learn. Boys are generally more active than girls. Boys are "diagnosed" as having ADHD four times more often than girls, simply because they are active. Nature has given boys and girls the means to learn about the world they live in. By exploring, investigating, and inquiring. Not by sitting for hours in the closed setting of a classroom. Nature's way is not convenient to traditional schools. The conjoined twins of the pharmaceutical and psychiatric industries saw an opportunity to brand the natural condition of child development as a disorder and make billions of dollars from the illusion.

"They (*psychiatry*) made a list of the most common symptoms of emotional discomfiture of children; those which bother teachers and parents most, and in a stroke that could not be more devoid of science or Hippocratic motive – termed them a 'disease'. Twenty-five years of research, not deserving of the term 'research', has failed to validate ADD/ADHD as a disease," emphatically states Dr. Fred A. Baughman, Jr. a pediatric neurologist. His book, *The ADHD Fraud*, clearly explains why parents should not believe anyone who wants to label their child as ADD or ADHD, even if that person has an M.D. behind their name.

Do you think the list below describes normal, healthy children less than 7 years old? My notes are in parentheses.

Acts quickly without thinking first. (The area of the frontal cortex that does analysis of situations to make good judgments is not fully formed until the early 20s.)

Cannot sit still. Walks, runs or climbs when others are seated. (Greek philosopher Aristotle had his students walk while he lectured because he realized movement enhances mental development.)

Daydreams or seems to be in another world. (Thomas Edison allowed time every day to daydream in order to be intuitively creative.)

Sidetracked by what is going on around him or her. (The survival of the human species depended upon awareness of what was going on in the environment. This is still true of individual survival. It is a natural process. Children learn to focus as they develop the skill over time.)

These are the APA voted-in indications of ADHD listed on the NIH website. How incredibly arrogant to think Nature no longer creates healthy children who are prepared to develop. Development takes time and varies for every child. Too often adults want instant results and they don't appreciate variation. Life must flow according to *their* time line. But life has its own rhythm and the children will pay the highest price for adults' impatience, arrogance, and greed.

How the ADD/ADHD Epidemic Began

The International Journal of Addictions lists more than **100 adverse reactions to methylphenidate and amphetamine drugs including: paranoid psychosis, terror, and paranoid delusion.** Adults are making children take drugs that literally give them living nightmares.

Where did it start? **In 1991 the Federal Education Department began to pay schools hundreds of dollars for every child diagnosed with ADD or ADHD.** It should not surprise anyone that suddenly there was an epidemic of ADD/ADHD in the U.S. And of course the way to treat the disorders is with drugs. The same drugs used in experiments to create extreme psychotic experiences in adult patients. Eight million children are made to take drugs that were never demonstrated to be beneficial and are listed by the DEA as addictive and as powerful as cocaine. And the drug companies profit. In 2008 U.S. sales of ADD/ADHD drugs were $4.8 billion.

If your child has suffered a side effect from an ADD/ADHD drug tell the FDA so their database will reflect the truth about these drugs. The instructions to report to the FDA are provided in chapter seven.

Children treated with methylphenidate (Ritalin) have a high drop-out rate and high incidence of drug and alcohol abuse, according to a study by Brookhaven Laboratory. They followed 5000 children made to take Ritalin. More than a third of the children did not complete high school and one-tenth attempted suicide. When the Ritalin-treated children reached adolescence they had a much higher rate of drug and alcohol abuse and a higher rate of criminal acts. There are additional studies showing similar results for children made to take amphetamines (Adderall and Dexedrine). The drug companies have tried to discredit the studies and have designed their own studies that conveniently make opposite claims, but with very few participants and for short timeframes.

An astounding 15.1 million Americans are getting high on ADD/ADHD drugs, other psychiatric drugs, and painkillers, as reported by the National Center on Addiction and Substance Abuse (CASA) at Columbia University in July 2005. Between 1992 and 2003 the number of children ages 12 to 17 abusing prescription drugs rose 212 percent. This astounding increase was absolutely predictable. Remember, the Federal

Education Department began to pay schools in 1991 for every child diagnosed with ADD/ADHD. Schools need money, they aren't thinking of the health of the children. The number of adults age 18 and older abusing prescription drugs increased 81 percent between 1992 and 2003. The U.S., with its constant focus on wealth, has allowed the pharmaceutical industry and psychiatric establishment to create drug addicts of a large percentage of an entire generation. The cost to society is immeasurable. The cost to the individuals may be their lives.

Children who abuse prescription drugs use illegal drugs at a much greater rate: 21 times more likely to use cocaine, 15 times more likely to use Ecstasy, and 12 times more likely to use heroin.

We cannot blame the children. It was not their choice to take mind-altering addictive drugs as young children.

Now the federal government is on the same road of helping the drug companies and psychiatry prey on children with money budgeted to go to states that screen teenagers and children for mental health. The treatment again is mind-altering, addictive drugs.

America please wake up!
If another country tried to drug our children
we would declare war!

Deadly Drugs

Matthew Smith died on March 21, 2002. He was 14 years old. The cause of death on the death certificate is "long term use of methylphenidate."(Ritalin) The chief pathologist said Matthew's heart showed clear signs of small vessel damage, the type of damage caused by stimulant drugs like amphetamines and methylphenidate. Matthew's heart weighed 402 grams at the time of his death. The heart weight of a healthy adult man is 350 grams.

Matthew's parents had never wanted him to take Ritalin. When Matthew was only 7 years old school officials told his parents they had diagnosed Matthew was ADHD. They threatened that if Matthew, a healthy active boy, was not put on Ritalin they would contact Social Service Child Protective Services. The school authorities told Matthew's parents they could be charged with neglecting Matthew's educational and emotional needs. His parents did not want Matthew on drugs and saw nothing abnormal about Matthew. But they also felt helpless to the school's authority and the fear that their child could be taken from them. They did what millions of parents across the U.S. did. They complied under duress. Now their beautiful, creative son is dead.

Stephanie Hall was only 11 years old when she died the day after her Ritalin dose was increased. Stephanie had been in first grade less than one month when her teacher told her mother Stephanie should be seen by the school psychologist because she may have ADD. Stephanie's parents took her to a different doctor instead. Everything was found as normal. Yet she was diagnosed with ADD and put on Ritalin. The drug gave Stephanie headaches. As Stephanie reacted to the Ritalin's adverse events the solution was always to increase the dose, even when she began to hallucinate. No one considered the drug as the cause.

That is the extremely dangerous fact about labeling someone with a mental disorder. **Once people are labeled with disorders, all their reactions will be blamed on the disorders, not the drugs.** In the case of psychiatric drugs, that is a deadly mistake. Stephanie's body kept compensating as best it could to each increase in Ritalin. Finally her body could not take it. She died in her sleep from cardiac arrest on January 5, 1996, six days before her twelfth birthday.

Stephanie and Matthew's deaths are not unique. In the ten years between 1990 and 2000 there were 186 children's deaths due to methylphenidate reported to the FDA. The FDA admits this number represents only 1 to 10 percent of the actual number of deaths, therefore the actual deaths are estimated between 1,860 and 11,860. Thousands of children have died in the U.S. for one reason – profit. The schools profit. The psychiatric profession profits. And the drug companies profit the most.

Score: Children – Zero, Drug Companies – Won

In February 2006 the FDA's Drug Safety and Risk Management Advisory Committee recommended all brands of drugs approved for ADD/ADHD have a Black Box Warning describing the cardiovascular risks. However, the FDA Pediatric Advisory Committee did not support the recommendation. **The FDA declined to include Black Box Warnings on these drugs.** Instead they instructed drug companies to add standard warnings for risk of heart attack and sudden death. After another year of deaths the FDA in February 2007 instructed all ADHD drug makers to add standard warnings for increased cardiovascular and psychiatric risks. Without a Black Box Warning, the FDA's most severe warning, on the package insert (PI) physicians and anyone checking the PI for adverse events will not be warned of how extreme the risk of cardiac arrest and sudden death is to children. The warning currently on ADHD drugs is:

"Hypertension and other Cardiovascular Conditions. Caution is indicated in treating patients whose underlying medical conditions might be compromised by increases in blood pressure or heart rate, eg, those with preexisting hypertension, heart failure, recent myocardial infarction (*heart attack*), or hyperthyroidism. Blood pressure should be monitored at appropriate intervals in patients taking *Drug Name*, especially patients with hypertension."

Stephanie and Matthew did not have preexisting hypertension nor had they any previous heart problems. The drugs caused their heart problems and caused massive heart attacks that did not give them a chance to survive. This warning does not give parents or physicians the high level of danger these drugs are to children. The FDA ruled against their Drug Safety and Risk Management Advisory Committee that had recommended a Black Box Warning.

No one warned Stephanie's or Matthew's parents of the risks with Ritalin. Not the addiction risks. Not the cardiovascular risks. Not the risk of sudden death. Prescriptions are handed out like they are for sugar pills. More children are going to die simply because of others' greed.

Canada suspended the use of Adderall in February 2005 after its maker, Shire Pharmaceuticals, reported 20 people including 12 children had died suddenly in the U.S. Dr. Robert Peterson, an official at Health Canada (Canada's FDA counterpart) said some of the deaths had occurred long before Health Canada approved Adderall XR in January 2004 but had not been reported by Shire. Canadian law allows regulators to remove a drug from the market while safety questions are investigated. U.S. law does not allow the FDA to withdraw drugs during safety investigations. Dr. Peterson said, "It's very difficult to generate a benefit-to-risk balance when the risk is sudden and unexpected death." Unfortunately the FDA does not share the same principles.

ADHD Drugs May Cause Cancer

All 12 children of a recent study displayed "significant treatment induced chromosomal aberrations" when they were given methylphenidate (Ritalin) standard therapeutic doses for only 3 months. The study, although small, was the first time ADHD drugs were tested to see if they are cancer causing. The results are significant because 100 percent of the children in the trial suffered chromosomal aberrations, indicating the drugs are toxic to human genes, therefore carcinogenic. The study was completed in February 2005 at the University of Texas M.D. Anderson Cancer Center. The study results won't disturb drug makers because people with cancer are another source of money for them. You see why drug companies are never concerned about drugs causing disease? Our drug-centered medical system's only answer is always another drug.

Deaths the FDA Won't Talk About

There are more deaths from ADD/ADHD drugs the FDA is covering up. Besides being highly addictive these drugs all interfere with the normal working of the mind. They cause hallucinations, delusions, and paranoia. These extremely dangerous drugs are what the APA and drug industry have convinced school systems and parents to feed to millions of American children daily.

Just as with SSRI/SNRI antidepressants, the FDA reports of deaths due to ADD/ADHD drugs have never included people killed by drugged children and adults. Such as Gerald and Marle McCra and their 11 year old daughter, Melanie. All were shot and killed by Gerald, Jr., age 15, who had been on Ritalin for nine years and suffered from paranoia.

Fifteen-year-old Rod Matthews lured fourteen-year-old Shaun Ouillette into a wooded area in 1986. Rod beat Shaun to death with a baseball bat. Rod had written in a journal a month before the murder, "My problem is I like to do crazy things. I've been lighting fires all over the place. Lately, I've been wanting to kill people I hate, and I've been wanting to light houses on fire. What should I do?"

Rod had been taking Ritalin since third grade. If someone had read his journal before the brutal murder would they have had the common sense to consider the drug as the cause of Rod's craziness? Not likely. Ritalin is for kids and it is FDA approved so it must be safe, right? Is cocaine safe? In the eyes of the DEA the two drugs are the same. Rod is now serving a life prison sentence.

Nicole Hadley, Jessica James and Kayce Steger's names are not included in the FDA Ritalin-related deaths. They should be. Michael Carneal, 14 years old, took guns into a school prayer meeting and began blasting away. Nicole, Jessica, and Kayce were killed and five more students were injured. One girl is permanently disabled. Michael was taking Ritalin. Friends told police Michael had been suffering from severe paranoia and unreasonable fears. He believed his family and friends were plotting against him. No one thought to have him stop taking Ritalin when these psychotic symptoms began.

Jeremy Strohmeyer was a high school honor student. Someone diagnosed him with ADHD and he was put on Dexedrine. Within days this young man with so much potential raped and murdered Sherrice Iverson, only 7 years old, in a casino on the California and Nevada border. Jeremy then tried to commit suicide and left a note asking for forgiveness. Jeremy lived but is serving a life sentence with no chance of parole.

Jeffrey Franklin was only 17 years old and had been prescribed the incredibly dangerous cocktail of three mind-altering drugs, Ritalin, Prozac and Klonopin. He bludgeoned and killed his parents with an ax and attempted to kill his sister and two brothers. Legal documents describe that Jeffrey told police he had crushed and snorted Ritalin and had not slept for days. He described having hallucinations and out of body experiences. Jeffrey is serving five life prison terms. Jeffrey and his family should be listed in the FDA adverse events records.

Kip Kinkel was diagnosed with dyslexia and put on Ritalin – an off-label use of the drug. Probably because of his reactions to Ritalin he was later prescribed Prozac also. Kip was taking two mind-altering, hallucinatory drugs when he took a gun to his school. He did not shoot anyone but was arrested at school, taken to a police station and charged. According to the detective who interviewed Kip, he was very concerned what his parents would think. His father bailed Kip out and they went home. According to Kip's confession his father was drinking coffee at the kitchen table later that day when Kip shot him in the back of the head, killing him. When Kip's mother arrived home from work Kip told her he loved her then proceeded to shoot her six times. He left a note on the coffee table, "I have just killed my parents! I don't know what is happening. I love my mom and dad so much. I just got two felonies on my record (*taking the gun to school*). My parents can't take that! It would destroy them. The embarrassment would be too much for them. They couldn't live with themselves. I'm so sorry. . .I can't eat. I can't sleep. (*adverse events for both drugs*) . . . My head just doesn't work right. God damn these VOICES inside my head."

The morning following his parents' murders Kip went to school with three guns and a backpack full of ammunition. He shot and killed two students and injured 25 students. One girl has bullet fragments permanently lodged in her brain causing brain damage. Kip was only 14 years old and had no idea why he had committed the diabolical acts. He was sentenced to 111 years in prison without chance of parole.

Dawn Branson had never suffered from psychosis. She was diagnosed ADHD and put on Adderall. Then as she was driving herself and her 3 year-old son across the Arizona desert a voice kept telling her, "Let go of the steering wheel and gas. God will drive the car. Don't you trust him?" Dawn finally did as instructed by the voice. The crash killed her son, Nathan.

Dawn was seriously injured. She quit taking Adderall and never heard voices again. The FDA doesn't list Nathan's death as ADHD drug related. They should.

There are hundreds of additional cases that are known deaths and injuries due to ADD/ADHD drugs. There may be thousands more that are unknown. Because these are legal drugs there is no official record keeping system. They all should be included in the FDA's system as important adverse event data so the complete picture of these drugs can finally be seen. By ignoring these deaths and injuries the illusion that these drugs are safe continues.

If these same tragic events had occurred while the attackers were taking illegal drugs, the authorities would immediately consider the drugs as the probable cause. **The line between legal and illegal is just an arbitrary designation**. Remember, LSD and Ecstasy have been legal drugs. The U.S. has been putting millions of young children on highly addictive, mind-altering drugs for more than two decades, FOR NO HEALTH REASON!

Smoke and Mirrors

"Methodologically rigorous research…indicates that ADD and hyperactivity as syndromes simply do not exist. We have invented a disease, given it medical sanction, and now must disown it. The major question is how we go about destroying the monster we have created. It is not easy to do this and still save face," writes education expert and author Diane McGuinness.

The list of APA voted-in indications for ADHD includes:

1. often fidgets with hands or feet or squirms in seat
2. often leaves seat in classroom or in other situations when remaining seated is expected

3. often runs about or climbs excessively in situations in which it is inappropriate
4. often has difficulty playing quietly
5. onset of symptoms before the age of 7

It is obvious that these describe the normal behavior of young, healthy children. With this sort of criteria, and the opportunity to gain hundreds of dollars for every child the schools diagnosed, it was absolutely predictable millions of children would suddenly be labeled with ADD or ADHD.

The age of children being prescribed these cocaine-like drugs continues to drop. Now that the schools are well saturated the preschools have been targeted by the "psychiatric-pharmaceutical cartel" as Dr. Fred Baughman, Jr. refers to it. Millions of children are given drugs that have never been proven to be beneficial. Does that make any sense? Nature can no longer create healthy children? The children are healthy - if they haven't been poisoned by heavy metals. It is the adults who drug children for profit who are sick.

> "When we drug millions of children to make them more compliant and easier to manage at home and in school, it says much more about our society's distorted values than about our children."
> Ginger Ross Breggin

Besides the fact that movement is normal in healthy children, there are many physical reasons why some children are especially active and inattentive.

"Artificial colors or sodium benzoate preservative (or both) in the diet results in increased hyperactivity in 3-year-old and 8 and 9-year-old children..." states a British study published in *The Lancet* on September 6, 2007. The study indicates the importance of providing children with natural foods. To feed children processed foods and then give them drugs because of the effects of the foods is unloving and unjust. We need to stop treating symptoms with drugs and get to the root causes of health issues.

Mary Stoddard's daughter experienced a grand mal seizure and lost consciousness on the playground. Physicians could not find a reason. Mary became suspicious of the chemical substitute for sugar, Aspartame. It is contained in NutraSweet, Equal, and Spoonful. Her daughter had started to have headaches when she began to drink beverages sweetened with Aspartame. When her daughter stopped consuming Aspartame she became healthy and never had any more seizures. Stoddard went on to investigate the dangers of Aspartame, commonly used in diet sodas and many sugar-free foods. She was the first non-MD to lecture at Southwest Medical School's continuing medical education series when she presented the truth about the health dangers of Aspartame and how easily and frequently Aspartame side effects were diagnosed as health disorders.

Aspartame causes symptoms that are frequently misdiagnosed as many diseases including bipolar, Multiple Sclerosis (MS), fibromyalgia, chronic fatigue syndrome, lupus, Lou Gehrig's disease, and Graves disease. Aspartame has also been clinically linked for decades with brain tumors.

More Heavy Metal Poisoning

A major reason for children to have trouble focusing is from lead poisoning. A 1996 study by Robert Tuthill found a significant relationship between the lead level in first-graders and the voted-in symptoms of the contrived ADD/ADHD.

Deborah Denno, Ph.D. published results of a study that followed 1000 children from birth to age 22. The results show a highly significant link between lead toxicity and the likelihood of criminal activity. The research found the best predictor of aggressive behavior in school, juvenile delinquency, and eventual criminal violence is the degree of lead poisoning. There have been several other studies confirming Dr. Denno's findings.

Twenty-five percent of all children in the U.S. are at risk for lead poisoning, according to a 2005 report by the American Academy of Pediatrics (AAP) . Because lead-based paint was allowed to be used until 1978, many homes and apartments in the U.S. remain a danger for lead poisoning. Lead paint does not have to be eaten to get into our bodies. Lead paint can produce lead dust that is inhaled. Lead can also be in water supplies from lead-soldered pipes.

"There is no known threshold below which adverse effects of lead do not occur and recent studies demonstrate that lead-associated intellectual deficits occur at blood lead levels less than 10 ug/dL," the AAP report states. In other words, there is no safe level for lead.

Lead causes a breakdown of the biochemical inhibition mechanisms in the area of the brain called the frontal lobe. One section of the brain stimulates excitement and impulsiveness. Another section of the brain inhibits actions so a person doesn't automatically react. In a healthy brain these two dynamics are balanced. In brains damaged by lead poisoning the balance is lost.

The methylphenidate and amphetamine drugs given to children only mask over some of the symptoms of lead poisoning. The drugs do not create balance in the two areas of the brain. They do nothing to help rid the body, also called detoxification, of lead or any heavy metal including mercury that causes neurological damage. And the combination of a brain damaged by lead and the addition of a mind-altering drug that increases the risk of violence sounds like a perfect prescription for violent, murderous acts.

Recently lead paint was found on toys made in China. Mattel Toy Company recalled their toys. But it is unknown how many thousands of toys by other companies have lead paint that is silently poisoning children around the world.

Beware of red lipstick. A study in 2007 found lead in all 33 different red lipsticks tested for lead. The lipsticks tested included expensive brands sold in department stores, brands sold in discount stores, and even a popular all-natural brand.

Children with lead poisoning need to be identified and provided detoxification methods to get the lead out of their bodies and ensure the environmental cause of their lead poisoning is removed. Identifying lead in children is a simple blood test. Environmental physicians are especially knowledgeable to help people with detoxification. Children should be tested for lead toxicity instead of being labeled with ADD/ADHD. There is no sense in poisoning a child twice, first with heavy metals then with cocaine-like drugs.

More Root Causes for Hyperactive or Inattentive Children

True healthcare is the process of diagnosing and treating actual causes of health problems. Western medical care is rooted in drug therapy that usually medicates symptoms, allowing health problems to continue.

Testing for food allergies and other allergies can uncover the reason some children fidget. Hearing testing is also important in children who have difficulty following directions. Dehydration can cause attention problems. Children's growing bodies need several cups of water every day. Milk and juice cannot replace the body's water needs. Only water can provide the necessary hydration that is imperative for healthy cells. Teaching your child to enjoy drinking pure, unadulterated water will put them on a healthy path for life. If you suspect lead is in your tap water, invest in a good filter that removes lead, or purchase tested spring water for drinking. Your family's health is worth it.

Well-nourished children who received daily essential minerals and vitamins performed significantly better on tests measuring their learning and memory capabilities, according to a study published October 2007. The study indicates that even children who are well fed benefit from daily supplements. The children in the study were given micronutrients containing the essential minerals iron, zinc, and folate; vitamins A, B-6, B-12 and C; and fish oils EPA and DHA. Until this study, there have been few studies assessing the impact that vitamin and mineral supplements have on the cognitive functioning of school children.

Allowing children to consume sugar and caffeine will definitely cause high levels of activity. Remember chocolate contains both, as do many soda beverages that are popular with children and adults.

ADD and ADHD are one-size-fits-all labels that conveniently generate huge profits for drug companies but do not provide healthcare to any children. Masking symptoms with mind-altering, addictive drugs is a terrible injustice.

Children need to be allowed to be physically active for extended periods every day. It is natural and required for good health. Activity level will naturally differ in children. Some healthy children will be at the higher end of the active spectrum. Hyper is in the eyes of the beholder. What is a hyper child to one adult may be a future Olympic athlete to another. Movement is healthy for humans of all ages. The 2006 study by Harvard University found physical inactivity as one of the seven leading negative factors of life expectancy. Encourage your children to turn off the computer and television and move. For their health - and no one's profit.

18

YOUR HEALTH IS YOUR LIFE

What is missing in this expensive, drug-centered medical environment? Why does the U.S. with the highest average per person healthcare spending rank 28th in the World Health Organization's Estimates of Healthy Life Expectancy (HALE)? If prescription drugs are the key to good health, why does the most drug-consuming nation by far, rank so poorly?

It is not about the amount of money spent. According to the 2006 United Nations Human Development Report the U.S. spent an average of $5,711 per person. Australia ranked 6th in HALE and spent an average $2,874 per person for healthcare. Canada ranked 17th and spent an average of $2,989. All figures are shown in U.S. dollars. Other sources listed U.S. per capita spending as high as $6,102 for 2006. Other countries use far less prescription drugs per person than the U.S., which consumes nearly half of the world's production of pharmaceuticals but has only about 5 percent of the world's population. Could the key to healthy living be something uncomplicated? Could it be easily accessible?

Life style behaviors are the leading contributors to life expectancy of Americans according to a Harvard School of Public Health study published in 2006. "There are millions of Americans who have life spans the same as in developing countries," says Dr. Christopher Murray, the lead researcher of the study. Americans are relying on the drug-centered medical system to fix health problems that are primarily self-inflicted.

Many physicians complained to me that patients ask for pills rather than make the effort to modify their behaviors. Most people do not want to make any change until a serious health problem arises. Then many still won't take responsibility for what happened. They just want the medical community to get them back to 100 percent.

You need to take responsibility for your health rather then relying on the quick fix of a pill. Life style changes can be made and do not have to cost money. We need physicians who understand the importance of whole-person healing and stop reaching for the prescription pad for every patient. The healthcare provided by the U.S. medical system will not improve until drugs are the alternative rather than primary remedy.

Life Expectancy is Based on Life Choices

The leading negative factors of U.S. life expectancy all point to life style choices, according to the Harvard study. The results of the study named the leading negative factors as tobacco, alcohol, obesity, high blood pressure, diet, and physical inactivity.

These factors are still a Western medicine point of view. Several are symptoms from unhealthy life choices. What needs to be addressed are the underlying root causes. Good health comes from addressing root causes, not from tampering with the symptoms. From a holistic health point of view, tobacco and alcohol are used by addictive personalities. According to Pat Love, Ph.D., author and authority on anxiety and depression, alcohol is often used to self-medicate for depression and/or anxiety.

Obesity, diet and often times high blood pressure are about eating behaviors and dehydration. Improved food choices and drinking at least 8 to 10 cups of water every day (for adults) will make a significant difference to any overweight person. Around the globe, but especially in the U.S., people are frequently consuming fast foods, processed foods, caffeine, and soft drinks. These foods and beverages contain too many unhealthy fats, high glycemic carbohydrates, and processed sugar or high fructose corn syrup (as healthy to consume as battery acid). The Academy Award-nominated documentary, "Super Size Me" by Morgan Spurlock dramatically demonstrates just how unhealthy fast foods are.

"There is no safe way of eating junk food, just as we learned the lesson from trans fats and partially hydrogenated oils often found in fat-free or low-fat cookies," explains Dr. Suzanne R. Steinbaum, director of Women and Heart Disease at Lenox Hill Hospital in New York City. "Diet soda does not protect us from the development of what we are trying to avoid by consuming it." Dr. Steinbaum was the lead researcher in a study that found people who consume more than one soft drink of any kind, regular or diet, daily are 44 percent more likely to develop metabolic syndrome (a precursor to diabetes) than people who do not drink any soda beverages. None of the 6,000 study participants had metabolic syndrome before the four-year study began. Metabolic syndrome is a collection of risk factors including high blood pressure, elevated levels of tryglycerides, low levels of HDL cholesterol (good cholesterol), high fasting blood sugar levels, and excessive waist circumference. Three or more of any of these factors indicates metabolic syndrome, and a much greater risk of developing cardiovascular disease and/or diabetes. The study was published in the medical journal *Circulation*, July 24, 2007.

Unfortunately many parents are teaching unhealthy life style choices to their children. Obesity and type-2 diabetes are occurring at younger ages. Have you ever offered your children or grandchildren a reward or bribe of eating at a fast food restaurant? Sure it is fast and easy for you. But you are putting your children onto a path of unhealthy eating that will lead to poor health. Make healthy choices for yourself and your children or grandchildren. Do you want them to become a statistic in the medical environment? Do you want your daughter or son to be diagnosed with diabetes? Diabetics have less than average life spans because diabetes initiates many other severe health problems.

Improving health requires dealing with the actual causes of poor health. But most people do not think they have poor health until something catastrophic happens. Many people are much more interested in the superficiality of looking good rather than striving to have optimal health.

Drugs Rarely Treat Causes

The drug-centered medical environment has not been good at addressing root causes. Instead it usually treats symptoms. For example, there is a prescription drug to reduce smokers' cough. The drug does not help a person stop smoking. It is a drug that pharmaceutical sales people encourage physicians to prescribe. If you woke in a smoke filled room would you look for a fan to blow the smoke out the window? Of course not. You know smoke is a symptom and the problem is a fire. Smoker's cough is not a health problem. The body is trying to tell the person, "Hey, you've got a problem here. I can't get oxygen. Stop smoking." Smoking is an addiction and requires the conviction to kick the addiction. But drug companies see an easy way to make money from all the addicts who do not want to stop smoking. Sell them a pill for the cough - a nice fast fix. Remind you of any other drugs?

How do you feel about your tax money being spent for this drug? As the study results show, smoking is one of the leading unhealthy behaviors. So while your tax dollars are buying pills to reduce the cough, the smoker's health problems are going to become more serious. The risk of cancer, stroke, and heart attack are all increased in people who use tobacco products. Any of these serious health incidents are extremely expensive. Your tax dollars will have to pay if it occurs in any of the 100 million people who receive medical treatment from government money. Before the government is completely bankrupted by our drug-centered medical system maybe it is time for some changes. Such as only paying for drugs for life threatening issues. Not paying for the drugs considered in the pharmaceutical industry as "life enhancing." Life enhancing? The side effects of many drugs are far from life enhancing. Some are life ending.

Integrative Medicine

"In many instances, knowing the person who has the disease is as important as knowing the disease that person has." J. McCormick, M.D. explains his medical view. This is a key missing point in traditional Western medicine. Physicians are trained to diagnose and treat disease. Reductionism is still the basis for diagnosing; look at the parts as though they are separate from the whole. The lack of a system approach to medical care can lead to additional health problems.

Whole-person medicine is just beginning in the U.S. The human form is a single system of interactive energy and information flow. When one area of the body-mind is in trouble the entire body-mind has a problem. When any type of treatment is given it will somehow affect the whole system. Western medicine treats diseases of the parts.

Neal Barnard, M.D. and author explains, "When I went to medical school, we learned a great deal about how to diagnose conditions, how to manage them medically, and how to prescribe drugs." Dr. Barnard's most recent book explains how to reverse diabetes without drugs through better eating.

Harvard trained physician, Dr. Andrew Weil's books about nutrition and plants that promote good health are excellent sources to improve your eating habits and gain a better understanding of health and aging. Dr. Weil expresses concern that many Americans are more interested in finding a fountain of youth. He said in a recent interview, "I consider the fixation on anti-aging and life extension to be a distraction from the important goal of healthy aging. That is, we should concentrate on making positive lifestyle choices now, eating better, exercising more, getting enough sleep, even improving our mental state, so that we can enjoy not just a longer life but a healthier one." Dr. Weil's book, *Eating Well For Optimum Health*, is an excellent resource to learn the truth about fats, carbohydrates and proteins. With this knowledge you can make healthier, yet desirable choices. Your eating habits will improve and so will your health. What have you got to lose? Maybe a few extra pounds.

His book, *Healthy Aging*, provides excellent information based on science about herbs and foods that help the body function optimally. For example the root of the Rhodiola rosea, also known as arctic root, golden root or rose root, has been used for hundreds of years in northern countries where it grows. Modern research has confirmed the root of the plant contains a group of compounds called rosavins that offer healthful properties. These include anti-stress, anti-fatigue, anti-cancer and immune system enhancement. It improves mood and memory. This is the type of information healthcare providers who practice holistic healthcare are familiar with. They may suggest a patient try arctic root from a health food store rather than reaching for a prescription pad.

Pharmaceutical companies will use their pervasive financial powers to create doubts and distrust about holistic healthcare. Because once a critical mass of the public realizes holistic care is better than drug-centered care, the drug companies know their control over the U.S. medical system is lost. That is another reason they now pursue mandatory legislation from the federal and state governments. It is an attempt to build a stronger fence around the American public, preventing an escape from drugs. But the decision is completely your own. You can say no to drugs as the first choice for your healthcare. Advocates are needed to convince insurance companies and third party payers to cover holistic healthcare so that people can afford care options. Healthcare should be a personal choice. After all, what is more personal than your health?

Energy Medicine

"**Energy is the currency of all interactions in nature.**" said Nobel laureate Dr. Albert Szent-Gyorgyi. His peers opposed his theory of semiconduction in living systems. His theory has since been shown to be entirely correct. The significance of semiconductors is that their conduction of electricity can be precisely controlled. Dr. Szent-Gyorgyi discovered that virtually all the molecules of the body are semiconductors. Dr. Szent-Gyorgyi accepted that people who think new thoughts are out of step with accepted philosophies. "Discovery is seeing what everybody else has seen, and thinking what nobody else has thought." He also said, "When everyone begins to laugh at you, then you know you are two steps ahead."

Albert Einstein identified in the 1920s that everything in the Universe, including the human form, is composed of energy. Dr. Szent-Gyorgyi's work identified the molecular component of the bioelectrical energy of living systems supporting Einstein's theory.

Western medicine completely ignores energy. There are no energetic concepts taught in physiology classes in U.S.

medical schools. "To leave energetic considerations out of the equations of life and medicine and healing is to ignore some 99 percent of what is happening. This is an enormous flaw in our medicine and our medical research," explains James L. Oschman, Ph.D. author of *Energy Medicine, The Scientific Basis.*

Dr. Oschman spent more than two decades researching the scientific findings supporting what the U.S. medical system considers complementary or alternative medicine (CAM). Because these findings are published in scientific journals not in traditional medical journals, medical professionals remain unaware of the vast amount of data explaining why acupuncture, therapeutic touch, sound therapy, light therapy and other forms of CAM do actually promote healing.

Some medical schools are beginning to address this void in medical education. Even the National Institute of Health created a department for research of CAM practices. As Dr. Oschman explains, "Subtle energies and dynamic energy systems are neither supernatural nor do they require a revision of physics. They go to the foundation of life."

What is new in the U.S. is actually ancient. All cultures except ours have used movement of energy for healing. It makes sense when you remember energy is the basis for all interactions in nature. Healing is a highly interactive process of energy and information movement within the human system.

Acupuncture has been successfully healing people in China for thousands of years. Light frequencies and sound vibrations have also been used for thousands of years in many cultures. Reiki, a form of healing touch, has been taught in Japan for many centuries. All these and many more are energy movement healing methods. The human body has electromagnetic fields. The strongest field is from the heart. Give yourself and others loving thoughts. It is healthy and healing. The second strongest field is from the hands.

"Molecules do not have to touch each other to interact. Energy can flow through...the electromagnetic field...The electromagnetic field, along with water, forms the matrix of life. Water...can form structures that transmit energy." Dr. Szent-Gyorgyi.

Your body is a living matrix where water plays a vital role in energy and information flow. "The importance of water cannot be over-estimated. Each fiber of the living matrix, both outside and inside cells and nuclei, is surrounded by an organized layer of water that can serve as a separate channel of communication and energy flow." Dr. Oschman explains in his book. Drinking good water every day is extremely important to good health.

Healing With Water

"You're not sick. You are thirsty. Don't treat thirst with medication." This was the sincere message the late Dr. Fereydoon Batmanghelidj (pronounced Batman-ge-lij) worked diligently to get out to the public and any healthcare providers who would accept his radical findings. Like any new discovery that opposes the established culture, Dr. Batmanghelidj's research has been ridiculed by the pharmaceutically-dominated medical establishment.

Dr. Batmanghelidj studied medicine with Sir Alexander Fleming, the Nobel laureate for his discovery of penicillin. Dr. Batmanghelidj went on to practice medicine first in his native land of Iran then in the United States. He discovered **many illnesses are not actually diseases but are the body's dehydration management.** The body is literally crying out for water. Dr. Batmanghelidj researched for more than 20 years the physiology of dehydration at the molecular level. How the body's systems react to conserve water is some of the most important information you can ever learn. It can literally save your life.

Because the concept that water can actually cure and prevent health disorders is so radical, most western physicians still choose to ignore all the scientific data. Dr. Batmanghelidj tried, without much success, for years to get the medical community to understand the seriousness of chronic dehydration and why drugs are not the answer. He then decided to write books for the general public so that people could address the root cause of many of their health problems, rather than treating the symptoms with drugs and never truly healing. Water cures many health disorders because the root cause of the problem is lack of sufficient water within the body-mind. The body-mind has a myriad of intricate ways to manage dehydration. You can be dehydrated and not feel thirsty. Dr. Batmanghelidj's books explain why water is so important and why western medicine has ignored the physiology of chronic dehydration. His website http://www.watercure.com provides several papers explaining the curative effects of water.

Some of the early indicators of dehydration include: tiredness, fuzzy thinking, heartburn, headaches including migraine headaches, lower back pain, joint pain, and fibromyalgic pain. Some of the common ailments that are actually indications the body is rationing water to areas because of chronic dehydration include: asthma, allergies, hypertension (high blood pressure), constipation, type 2 diabetes, and chronic tiredness or depression.

A healthy adult body is 75 percent water and the brain is 85 percent water. The body-mind will try to tell you in many ways it is in need of life-giving water. Western medicine focuses on the solid matter and ignores the extremely important fact that water provides the energy for all functions of the body-mind.

Dr. Batmanghelidj's research findings tie directly in with the discoveries of Dr. Szent-Gyorgyi. The molecules of the body are semiconductors and water is the key element required for molecular energy transfer. As the body's ability for energy transfer degenerates, health is affected.

The shift in awareness from a view of the body as solid parts to the view of water's essential role is like the shift from thinking the world is flat to realizing the world is round. There really isn't any solid matter. All body-mind matter is comprised of water because energy transfer is the essence of life and water is vital to conducting energy.

"In any form of stress, you should immediately drink copious amounts of water. This is why dehydration stresses your body-mind, and stress precipitates so many disease conditions in the body." Dr. Batmanghelidj explains why drinking water is vitally important and not to wait until you feel thirsty. It is very important for elderly people to drink 8 to 10 cups of water every day. Studies have shown chronic dehydration leads to the body losing the ability to recognize thirst. Elderly people are healthier when they consume enough water daily.

The body loses an entire quart (32 ounces or 4 cups) of water every 24 hours just from normal breathing. The cloud seen when people exhale in cold weather is water released, and it occurs with every breath in all weather. Then there is sweat, urination, solid waste passing, and food digestion. All daily bodily functions that require water in addition to the maintenance of 75 percent water throughout the body and 85 percent in the brain.

Children are in a continual state of dehydration because growing bodies use more water. A few sips of water at the school water fountain are insufficient to address the water needs of a child. Milk and fruit juices do not compensate for water. Encourage your child to drink at least 8 ounces of water when they get up in the morning and as soon as they get home from school. Children with asthma are children with bodies screaming for water. Dr. Batmanghelidj's books are a good resource to learn about giving children and babies proper water and trace mineral amounts.

There are physicians who treated their own health problems with proper daily hydration, after discovering Dr. Batmanghelidj's research. Their letters of testimonials to Dr. Batmanghelidj are included in his books. Letters of gratitude from many people who thought they were diseased and would have to take drugs the rest of their lives, indicate the seriousness of chronic dehydration.

This knowledge could save your life if you are one of the millions of people with health disorders that are actually the body's call for more water. Treating the signals of dehydration with drugs allows your body to remain in a chronic dehydrated state that will continue to worsen. If you want to be healthy, drink water. If you want healthy aging, drink water. Just as the planet cannot support life without water, your body will not continue to live without proper water amounts.

The drug companies have waged campaigns to discredit water as the cure for so many ills.

Consider who makes drugs.
Consider who made water.
To whom do you trust your health?

More Easy Health Ideas

Thoughts are energy. Thought energy is the energetic basis behind the popular video "The Secret". You have heard of brain waves. These continuously occurring energy waves alternate at varying frequencies. That is why hypnosis has been used successfully in medicine for more than 200 years. Even the Mayo brothers, founders of the world renowned Mayo Clinic, used hypnosis in order to use minimal amounts of dangerous general anesthesia during surgeries.

The healing power of hypnosis has been underutilized since it went on stage as entertainment, creating many misconceptions. Hypnosis is not about someone else controlling you. Studies have shown people will simply come out of trance if they are directed to do something they do not want to do. Hypnotic trance gives you the ability to have extremely clear communication within yourself by quieting the judgmental mind. It can be far more effective than drugs without any side effects. The death rate of trauma victims was significantly reduced when emergency medical technicians were taught to use hypnosis at the scene of accidents. Hypnosis helps the communication within the body-mind to focus, and focus of energy helps the healing process.

A leading edge healthcare field is Energy Psychology. People are being helped quickly with grief, trauma, anxiety and depression with techniques that utilize the energy fields throughout the human form. Bruce Lipton, Ph.D. author of *Biology of Belief* explains, "Conventional physics sees the human body as a machine made of atoms and molecules but the quantum physicists reveal that underneath that apparent physical structure there is nothing other than energy... What we are beginning to recognize is that there is an invisible world that we have not dealt with in regard to understanding the nature of our health. In other words, rather than focusing on matter, in a quantum world we focus on energy."

Light and Dark Therapy

Light therapy is well established as a treatment for seasonal affective disorder (SAD), a form of depression that occurs in winter months when there are fewer daylight hours. Full spectrum lights help people with SAD by simulating sunlight.

Dr. Jim Phelps believes **dark therapy is one of the easiest methods for someone with bipolar disorder.** A recent study on the effects of dark therapy for bipolar (manic-depressive) disorder showed promising results. Researchers at the Corvallis Psychiatric Clinic in Corvallis, Oregon studied what they are calling "virtual darkness." They created artificial darkness using amber lenses that block out blue spectrum lights (florescent, incandescent and LED). This produced the effects of total darkness while allowing the bipolar patients to see and continue evening activities. The study found bipolar patients were helped with 14 hours of the total darkness; six hours of created-by-lens darkness and eight hours of sleep. There are no side effects from light therapy.

Naturopathic physician Dr. Emily Kane has been using green light therapy for patients who are stressed. The green light frequency reduces the stress hormone cortisol. She explains that 30 to 45 minutes of green light daily is effective as a natural stress buster. Dr. Kane suggests doing evening activities like reading by green light to improve sleep.

Sunlight Provides Essential Vitamin D

Vitamin D3 reduces risk of all cancers by 77 percent in postmenopausal women, a study reported in the June 2007 *American Journal of Clinical Nutrition*. The 4-year study by Creighton University School of Medicine included almost 1200 healthy women age 55 or older. The group of women taking 1100 IU of vitamin D3 and 1400-1500 mg of calcium had a 77 percent reduced risk of all cancers. The group taking just the calcium supplement did not have any improved cancer benefits than the group taking the placebo. Vitamin D supplements are available in two forms, vitamin D2 and vitamin D3. The Creighton researchers recommend vitamin D3 because it is more active.

People living in southern latitudes have lower cancer rates than people living in northern latitudes where there is decreased sunlight. Humans make their own vitamin D3 when exposed to sunlight. Only 10 to 15 minutes of sun exposure in summer without sunscreen provides a sufficient daily dose for light skinned people. Dark skinned people need longer sunlight exposure. The rate of breast cancer in dark skinned women is greater in northern states where vitamin D deficiency is prevalent.

The Canadian Cancer Society now recommends 1,000 IU vitamin D supplements daily for adult Canadians during the fall and winter. "Where a person lives is one important factor in how much vitamin D they can produce from the sun. Because of our country's northern latitude, the sun's rays are weaker in the fall and winter and Canadians don't produce enough vitamin D from sunlight during this time." Heather Logan of the Canadian Cancer Society explains. The Society also recommends year-round vitamin D supplements for Canadians with dark skin, those who do not get outside often, and those who wear clothing that covers most of their skin.

Vitamin D is important in preventing many health disorders including osteoporosis. Vitamin D is vital to the process of calcium deposition in bones.

The occurrence of pre-eclampsia, a major contributor to complications during pregnancy and postpartum periods, was five times higher in women with low vitamin D levels during early pregnancy. Even when women take prenatal vitamins they are at high risk of insufficient vitamin D because the international units (IU) in most vitamin tablets is insufficient.

According to *The New England Journal of Medicine*, there is high risk of developing autoimmune diseases such as type-one diabetes, thyroid problems, asthma and low birth weight due to low levels of vitamin D. The current recommended daily allowances (RDA) for vitamin D are too low, according to recent scientific evidence. The Creighton research used 1100 IU of vitamin D3 daily which provided the 77 percent reduction of risk for all cancers. The RDA for the age group of women in the study, ages 50 to 70, is only 400 IU, an insufficient amount.

Laughter is Great Medicine

Enjoyment of life is a key to good health. Laughter releases endorphins giving us a sense of well-being and boosting our immune systems. Of course there are going to be times of sadness and grief. Remind yourself they are temporary, "This too shall pass." Envision a positive future. It is chronic negativism that stresses the human system. Have an attitude of gratitude. Having gratitude for the blessings you have instead of wishing for what you do not have is another important (and free) source to healthy living.

It is a philosophy promoted by popular singer Naomi Judd since her potentially fatal struggle with hepatitis C virus. "These guys with starched white coats and degrees told me I was going to die. If I had believed them, I'd be dead by now." Judd explained in a recent interview. She used integrative (CAM) treatments including biofeedback, aromatherapy, and meditation. Her hepatitis C is now in remission and she is living a fulfilling life as a committed crusader for raising awareness of the disease which will kill four times as many Americans as AIDS in the next 20 years. Judd's simple advice is: "Be optimistic. Feel good about wisdom gained. Be grateful, because you will want less. Stay close to the ones you love. Heal wounds. Let go."

More Simple Means to Good Health

Get a pet. Several studies have found dog owners go to doctors less often, have fewer illnesses, and recover more quickly when they do get sick. Petting an animal has been shown to lower blood pressure (beats taking pills) and is a marvelous antidepressant. The only side effect may be dog drool. If you positively cannot own a pet, consider volunteering at a local animal shelter. There you will find limitless amounts of unconditional love.

Sleep your way to good health. Many research findings have shown sleep promotes good health and can add years of healthy living. Before you reach for a prescription drug to help you sleep, try to find ways to relax and promote good sleep naturally. Aromatherapy with sleep inducing scents like lavender, vanilla, or green apple can be helpful. Reducing the amount of light before you want to sleep may help you sleep better. Bright light indicates to your body that you need to be alert.

Andrew Weil, M.D. suggests taking calcium citrate before bedtime. Calcium promotes sleep and is important to the heart's bioelectric energy cycles. You could also get the benefit of calcium by eating organic yogurt before going to bed.

If you sometimes use antihistamines to help you sleep, remember they cause additional dehydration. That is why you may be groggy when you wake. Drink water immediately when you get up every morning. The body loses one to two cups of water just from normal breathing while you sleep.

Try listening to relaxing music, natural sound tapes, or hypnosis tapes as you go to sleep.

Avoid stress before bed. Do not watch the news before bed. The negativity and violence are sleep disrupters for many people. Take care of any activities that may cause you stress such as balancing your checkbook or paying bills early in the day, not before bedtime.

If you consistently feel tired in the morning, you may suffer from sleep apnea. You should see your physician or go to a sleep disorder clinic to diagnose sleep apnea. A common cure for sleep apnea sufferers is oxygen, not drugs.

Insufficient sleep, less than seven or eight hours per night, can lead to weight gain, heart problems, even optical nerve damage. If you care about your health, get the rest your body needs for good health.

Reducing Stress

During stressful times breath awareness practices can prevent the release of steroids and other stress hormones that are very damaging to your health. One calming breathing technique is to count to seven as you breathe in deeply, allowing your abdomen to expand. Then hold the breath while you count to four. Exhale to a count of seven, getting all the air out. Then count to four before inhaling. Repeat for three or four total deep breaths. Breath awareness is the fastest way to rebalance when negative emotions begin to take over. What could be better than healthcare that is easy and free? Breath awareness can easily be used in the workplace where people are often stressed.

Another excellent method to rebalance yourself when anxiety, depression, fear, grief, anger, or any negative emotion overwhelms you is the Emotional Freedom Technique (EFT) created by Gary Craig. His website is www.emofree.com. You can use Gary's easy techniques anytime because they are patterns of tapping energy points on yourself. Psychologists and counselors are finding this technique offers fast, effective results.

Tom Thomson of Southern Pines, N.C., a professional counselor, has had people come to him with physical pain that physicians have not been able to diagnose. The EFT tapping technique eliminated the pain in a few minutes because the pain was caused by impaired energy flow due to stress. Conventional medicine is unaware of using energy movement for healing. Energy healing is not supernatural. It is the science of energy as Einstein, Szent-Gyorgyi, Becker, Frohlich and many other brilliant scientists have demonstrated in their work.

Meditation, yoga, and tai-chi have all been shown to have health benefits. One reason is because they strengthen the white blood cells known as memory T cells. These cells fight recurrence of previously experienced infections. Some infections can lie dormant in your body for years, such as the virus that causes first chickenpox then shingles. We lose memory T cells as we age. Doing activities such as meditation, yoga and tai-chi boosts the functioning of the memory T cells that you have.

Yoga and tai-chi also help maintain flexibility which improves blood flow, promotes good balance and helps maintain good skeletal structure. Sound skeletal structure and good balance help prevent falls. There is good reason that these ancient practices have continued to be used for centuries - they work.

Social interactions are important to your health. Having close friends or family members with whom you can interact on a regular basis promotes good health. It is important to feel you have others you can rely on and to share good times. Isolationism is not healthy, yet in our ever-busy life style we often become isolated. Instead of gulping down ineffective, dangerous antidepressant drugs, people would be healthier by finding ways to create community. Something as simple as sharing a potluck dinner with friends can be a good stress releaser. Never mind that your home isn't perfect. Relaxing with friends is more important than dusting. Sharing happy times with others at least once a week is a reasonable goal to set for your health. Turn off the television and computer. Make the effort to connect with people you care about. It will benefit your health and theirs.

Turn off the television and go for a walk. This simple free activity will help in multiple ways. It will get you away from the omnipresent drug advertising. It will get you away from the negativism on the news and many television shows. You will be outside in natural light that has proven health benefits. And you will get physical exercise, a leading factor in life expectancy. Multiply the health benefits by walking with a friend.

Natural Cures for Depression

"Don't think, do" is the mantra of a depression curing program created by psychologist Stephen Ilardi, Ph.D. of the University of Kansas. "There's increasing evidence that we were never designed for our sedentary, socially isolated, indoor, sleep-deprived, poorly nourished lifestyle, " Dr. Ilardi explains. He created a regimen called Therapeutic Lifestyle Change for Depression (TLC). The program includes a combination of long-proven depression remedies: increased sleep, aerobic exercise, daily doses of 1,000 milligrams of omega-3 supplement, exposure to bright, natural light, social interaction, and activity rather than rumination – don't think, do. People are given training in techniques of how to cease rumination. Group therapy is also included. TLC is a holistic approach to improving health, treating the body-mind-spirit as a system.

In 14 week-trials the TLC program has an impressive 76.6 percent positive response compared with the control group that was on antidepressant drugs and/or traditional psychotherapy having only 27.3 percent favorable results. The placebo group that received nothing had nearly identical results as the group receiving drugs, but without the dangerous side effects and addiction.

Dr. Ilardi's theories are supported by the studies of anthropologist Edward Schieffelin, who found contemporary hunter-gathering tribes in remote areas of the world are virtually depression free. Amish communities have also been studied and their outdoor activity-based and community-oriented lifestyle results in extremely low rates of depression.

Women's Depression Linked to Low Progesterone Levels

Lack of progesterone is the most likely reason for the uncomfortable symptoms of premenopause and menopause. The late John Lee, M.D. was the first to recognize that insufficient progesterone to balance estrogen was the real cause for the variety of discomforts from menopause. While Wyeth was paying millions of dollars for marketing to have women and doctors believe estrogen deficiency was the cause, Dr. Lee was quietly getting to the truth and writing about it in his book, *What Your Doctor May Not Tell You About Menopause.*

The reason middle-aged women often thicken around their midsection is because progesterone is stored in body fat. As menopause begins, production of progesterone drops at a much faster rate than estrogen. The body adds fat around the middle as a means to store as much progesterone as possible. Chronic stress at any age will deplete the body of progesterone.

Dr. Lee discovered women's ratio of estrogen to progesterone is the basis for many normal menopause symptoms and other health issues. He was the first to warn of the impact of environmental contaminants that mimic estrogen in both women's and men's bodies. These chemicals are in many plastics including plastic bottles contaminating liquids and foods. These environmental xenoestrogens, as they are called, are recognized as the reason girls in the U.S. and other western countries are physically developing at much younger ages.

When the body lacks sufficient progesterone to balance estrogen, severe symptoms can occur including: anxiety, depression, breast tenderness, fibroids, cyclical headaches or migraines, mental fatigue, irregular bleeding, water retention, and weight gain. These represent the most common. Some women may have other symptoms due to the hormonal imbalance. These are also symptoms that have been APA voted-in as characteristics of the contrived premenstrual dysphoric disorder (PMDD) for which drug companies push doctors to prescribe antidepressants. SSRI/SNRI drugs do not balance women's hormones. If you are a woman with these types of monthly symptoms you may find relief with natural progesterone.

Estrogen causes release of cortisol, a stress hormone. Progesterone neutralizes cortisol. Women suffering with anxiety or depression may want to consider trying natural progesterone. It is sold as creams over-the-counter. Because stress and aging deplete the body of progesterone, it is possible the primary cause of anxiety and depression in many women is due to lack of sufficient progesterone.

There are also compounded natural hormones. If your city has a compound pharmacy, you can talk to your physician about taking Progest, a natural progesterone that comes from the same vegetable as the cream progesterone. There is also a natural estrogen, Biest. A compound pharmacy, as mentioned earlier, can individualize the dose to your needs. This allows flexibility if circumstances cause life to become more stressful, you can personally modify your daily progesterone amount. Estrogen and progesterone levels vary throughout the day, so a one-time hormone test is not an accurate reading of your hormone levels.

The advantages of the natural progesterone cream compared to the compounded natural progesterone is that no prescription is necessary, and by applying it to your skin you receive much greater effect because the progesterone is not processed through the liver like ingested progesterone is. The liver removes up to 90 percent of the hormone.

Natural hormones are entirely different from the hormone replacement drugs Provera, Prempro, and Premarin. As author Joan Borysenko, Ph.D. describes in her book, *A Woman's Book of Life*, "The synthetic progestins, like Provera, that are included in most HRT prescriptions do not have the same biological effects as natural progesterone and can create a host of side effects...Natural progesterone, on the other hand, has no known side effects. It is not patentable and therefore of little interest to drug companies." Even though natural hormones come from plants (yams) they are chemically identical to human hormones. That is why they are called bioidentical. Synthetic hormones are not bioidentical.

If you admire horses there is another reason you do not want to take synthetic hormones. These hormones are made from the urine of pregnant mares. Mares have 17 different estrogens. These are not natural hormones for humans and are not bioidentical to human estrogen. The horses are kept under extremely inhumane conditions in order to collect the urine. Pregnant mares are permanently catheterized and confined in small stalls. After they give birth the foal is only allowed to nurse for one week so the mare can be impregnated again. The mares often die during the second or third pregnancy because of the unnatural and stressful conditions. If you are taking any of the synthetic HRT mentioned above, consider speaking with your physician about changing to natural hormones. Do it for your health and for the horses. Synthetic estrogen is one of the fifteen leading drugs for serious side effects reported to the FDA between 1998 and 2005.

Healthcare is up to us. We cannot change the medical environment. We can make changes to improve our individual health. If you are not yet convinced to take healthy steps, I will tell you something that may help to motivate you.

The Medical System is a Leading Cause of Death in the U.S.

System as defined by Webster is "a set or arrangement of things so related or connected as to form a unity or organic whole". That is how the human form functions. It is a system intricately connected by energy and information to form an organic whole. The medical care currently in the U.S. does not fit Webster's definition. We have a medical muddle, a "confused or disordered condition" (Webster). That medical muddle and the more than 106,000 deaths due to drug adverse events are **responsible for 225,000 to 250,000 deaths a year**. Making the medical system the third leading cause of death in the U.S. This was the conclusion of a study done by Dr. Barbara Starfield of the Johns Hopkins School of Hygiene and Public Health. The muddle creates unnecessary surgeries, medication errors in hospitals, other errors in hospitals, and infections in hospitals.

Then there are the injuries and deaths in hospitals due to drugs. A 2006 report, *Preventing Medication Errors*, by the Institute of Medicine states that **1.5 million people are injured or die in hospitals annually from drug errors including off-label use of drugs.**

There are some researchers who contend the medical muddle is the number one leading cause of death in the U.S. because the Johns Hopkins research only looked at hospital deaths and drug adverse event deaths. A review of medical peer-review journals and government health statistics compiled by several researchers in 2004 disclosed more than 780,000 deaths due to medical errors, unnecessary procedures, hospital infections, and adverse drug reactions. The death toll due to legal drugs and all the different ways that things can go wrong in the medical muddle may far surpass the deaths due to heart disease, which has been named the leading cause of death in the U.S. for decades. No matter which study results you accept, it is in your best interest to stay healthy and stay out of the muddle.

If you don't pay attention to being healthy, you will pay attention to being sick.

In their book, Critical Condition – How Healthcare in America Became Big Business and Bad Medicine, Donald L. Barlett and James B. Steele provide ample evidence of just how unsystematic the medical muddle is. Their book also explains where so much of the billions of dollars of healthcare money are actually going. In the years since the book was published conditions have worsened. We cannot hope for a complete transformation of this jumble of self-interested parties vying for more of the medical money that is now in the trillions of dollars. Do you want to be one of the statistics of this leading cause of death? If not, you need to do everything you can to remain out of the muddle. And keep your children out of the muddle. And keep your elderly relatives out of the muddle. Help keep anyone you care about out of the muddle.

Take Responsibility For Your Health

You need to take the responsibility for your health and make healthy life choices now before you are a victim of the medical muddle. Do not depend on a pill to fix everything you don't like about yourself or your life. Make a commitment to yourself to make healthy life choices while you have good health. People who stop smoking after a heart attack are already living with a damaged heart and a journey through the medical muddle; if they survived the heart attack and the muddle.

You can make prescription drugs a last resort, not the first quick fix because now you appreciate that in the majority of cases, drugs are used to treat symptoms, not the root cause of health problems. Try to find naturopathic physicians in your area or physicians who practice integrative medicine or whole-person medicine. Or when your physician reaches for the prescription pad ask him/her if life style changes could help. Don't be afraid to tell your physician you want to use drugs as a last resort. Remember that often the older, generic drugs are just as effective and may be safer than newer, more expensive drugs. Always ask for and read the package insert for any drug you are prescribed. Use the power of the Internet to investigate natural remedies or drug information. Protect your health. It is your life.

Some pharmaceutical drugs, such as antibiotics, provide appropriate care. But for many health conditions, drugs are not the answer. No matter what the drug companies would have the public believe, drugs do not belong in everyone's medicine cabinet. The drug-centered medical muddle has lulled Americans into the false idea that drugs are the best way to fix any health issue. I hope this book has helped you to realize the dangers of that myth.

EPILOGUE

February 2008

"Ms. Carlson, I am a lobbyist in Phoenix. There is a bill before our legislature that will require pharmaceutical companies to declare all gifts to doctors. We would like to help this bill get passed. Would you be willing to speak before the committee?"

I quickly agreed and plans were made for my transportation from my friend's home in Phoenix where I was visiting. I prepared just as I used to prepare to give seminars, with notes of important points to make.

"The pharmaceutical lobbyists are coming out of the woodwork to fight this bill even getting to vote," Richard explained as we drove to the Arizona State Capital on a sunny, warm day. After leaving the car in a public lot, he guided me into a beautiful building and down a long corridor to the room assigned to the committee.

"We should be scheduled for ten o'clock," he said looking over a posted sheet. But we were not on the schedule for ten or any other time. "Maybe there was a delay of other issues," he said with a frown. He quietly opened the door to the committee room. There was no one in the room. "This is strange," he said. "Wait here and I'll try to find out what is going on," he said turning and walking back to the main lobby.

A few minutes later he returned with an exasperated expression. "The pharmaceutical lobbyists got to the committee chairman and had your presentation canceled. They have him in their pocket. I'm really sorry to have put you to this trouble."

"You mean the committee chairperson has the power to decide if the committee will hear from a whistleblower?" I asked.

"That's right. That is why it is so easy for the pharmaceutical industry to maintain control. They financially support the key people like this chairman," Richard explained. "While we are down here I'll introduce you to some Representatives who are trying to hear the truth about the pharmaceutical industry."

We spent the morning meeting and talking with busy Representatives. I was grateful for the experience because I had been unaware of how easy it was for wealthy industries like pharmaceuticals to control governmental procedures. I became all the more determined to make this book available to the public.

Appendix A

Antidepressant Drugs Approved in the U.S.

All With Black Box Warnings for Risk of Suicide
For People Under the Age of 25

Anafranil (clomipramine)
Asendin (amoxapine)
Aventyl (nortriptyline)
Celexa (citalopram hydrobromide)
Cymbalta (duloxetine)
Desyrel (trazodone HCl)
Elavil (amitriptyline)
Effexor (venlafaxine HCl)
Emsam (selegiline)
Etrafon (perphenazine/amitriptyline)
Lexapro (escitalopram oxalate)
Limbitrol (chlordiazepoxide/amitriptyline)
Ludiomil (maprotiline)
Lovux (fluvoxamine maleate)
Marplan (isocarboxazid)
Nardil (phenelzine sulfate)
Pamelor (nortriptyline)
Parnate (tranylcypromine sulfate)
Paxil (paroxetine HCl)
Pexeva (paroxetine mesylate)
Prozac (fluoxetine HCl)
Remeron (mirtazapine)
Sarafem (fluoxetine HCl)
Seroquel (quetiapine)
Sinequan (doxepin)
Surmontil (trimipramine)
Symbyax (olanzapine/fluoxetine)
Tofranil (imipramine)

Tofranil-PM (imipramine pamoate)
Triavil (perphenazine/amitriptyline)
Vivactil (protriptyline)
Wellbutrin (bupropion HCI)
Zoloft (sertraline HCI)

Appendix B

H. R. 2218 Parental Consent Bill of 2009

(1) The United States Preventive Services Task Force (USPSTF) issued findings and recommendation against screening for suicide that corroborate those of the Canadian Preventive Services Task Force. "USPSTF found no evidence that screening for suicide risk reduces suicide attempts or mortality. There is limited evidence on the accuracy of screening tools to identify suicide risk in the primary care setting, including tools to identify those at high risk."

(2) The 1999 Surgeon General's report on mental health admitted the serious conflicts in the medical literature regarding the definitions of mental health and mental illness when it said, "In other words, what it means to be mentally healthy is subject to many different interpretations that are rooted in value judgments that may vary across cultures. The challenge of defining mental health has stalled the development of programs to foster mental health (Secker, 1998)."

(3) A 2005 report by the National Center for Infant and Early Childhood Health Policy admitted, with respect to the psychiatric screening of children from birth to age 5, the following: "We have mentioned a number of the problems for the new field of IMH (Infant Mental Health) throughout this paper, and many of them complicate examining outcomes." Briefly, such problems include:

(A) Lack of baseline.

(B) Lack of agreement about diagnosis.

(C) Criteria for referrals or acceptance into services are not always well defined.

(D) Lack of longitudinal outcome studies.

(E) Appropriate assessment and treatment requites multiple informants involved with the young child: parents, clinicians, child care staff, preschool staff, medical personnel, and other service providers.

(F) Broad parameters for determining socioemotional outcomes are not clearly defined, although much attention is being given to school readiness.

(4) Authors of the bible of psychiatric diagnosis, the Diagnostic and Statistical Manual, admit that the diagnostic criteria for mental illness are vague, saying: "DSM-IV criteria remain a consensus without clear empirical data supporting the number of items required for the diagnosis. . . Furthermore, the behavioral characteristics specified in DSM-IV, despite efforts to standardize them, remain subjective. . ." (American Psychiatric Association Committee on the Diagnostic and Statistical Manual (DSM-IV 1994), pp. 1162-1163).

(5) Because of the subjectivity of psychiatric diagnosis, it is all too easy for a psychiatrist to label a person's disagreement with the psychiatrist's political beliefs a mental disorder.

(6) Efforts are underway to add a diagnosis of "extreme intolerance" to the Diagnostic and Statistical Manual. Prisoners in California State penal system judged to have this extreme intolerance based on race or sexual orientation are considered to be delusional and are being medicated with antipsychotic drugs. (Washington Post 12/10/05)

(7) At least one federally-funded school violence prevention program has suggested that a child who shares his or her parent's traditional values may be likely to instigate school violence.

(8) Despite many statements in the popular press and by groups promoting the psychiatric labeling and medication of children, that ADD/ADHD is due to a chemical imbalance in the brain, the 1998 National Institutes of Health Consensus Conference

said, ". . .further research is necessary to firmly establish ADHD as a brain disorder. This is not unique to ADHD, but applies as well to most psychiatric disorders, including disabling diseases such as schizophrenia . . . Although an independent diagnostic test for ADHD does not exist. Finally, after years of clinical research and experience with ADHD, our knowledge about the cause or causes of ADHD remains speculative."

(9) There has been a precipitous increase in the prescription rates of psychiatric drugs in children.

(A) The use of antipsychotic medication in children has increased nearly fivefold between 1995 and 2002 with more than 2.5 million children receiving these medications, the youngest being 18 months old. (Vanderbilt University, 2006)

(B) More than 2.2 million children are receiving more than one psychotropic drug at one time with no scientific evidence of safety or effectiveness. (Medco Health Solutions, 2006)

(C) More money was spent on psychiatric drugs for children than on antibiotics or asthma medication in 2004. (Medco Trends, 2004)

(10) A September 2004 Food and Drug Administration hearing found that more than two-thirds of studies of antidepressants given to depressed children showed that they were no more effective than placebo, or sugar pills, and that only the positive trials were published by the pharmaceutical industry. The lack of effectiveness of antidepressants has been known by the Food and Drug Administration since at least 2000 when, according to the Food and Drug Administration Background Comments on Pediatric Depression, Robert Temple of the Food and Drug Administration Office of Drug Evaluation acknowledged the "preponderance of negative studies of antidepressants in pediatric populations." The Surgeon General's report said of

stimulant medication like Ritalin, "However, psychostimulants do not appear to achieve long-term changes in outcomes such as peer relationships, social or academic skills, or school achievement."

(11) The Food and Drug Administration finally acknowledged by issuing its most severe Black Box Warnings in September 2004, that the newer antidepressants are related to suicidal thoughts and actions in children and that this data was hidden for years. A confirmatory review of that data published in 2006 by Columbia University's department of psychiatry, which is also the originator of the TeenScreen instrument, found that "in children and adolescents (aged 6-18 years), antidepressant drug treatment was significantly associated with suicide attempts. . .and suicide deaths. . . " The Food and Drug Administration had over 2000 reports of completed suicides from 1987 to 1995 for the drug Prozac alone, which by the agency's own calculations represent but a fraction of the suicides. Prozac is the only such drug approved by the Food and Drug Administration for use in children.

(12) Other possible side effects of psychiatric medication used in children include mania, violence, dependence, weight gain, and insomnia from the newer antidepressants; cardiac toxicity including lethal arrhythmias from the older antidepressants; growth suppression, psychosis, and violence from stimulants; and diabetes from the newer antipsychotic medications.

(13) Parents are already being coerced to put their children on psychiatric medications and some children are dying because of it. Universal or mandatory mental health screening and the accompanying treatments recommended by the President's New Freedom Commission on Mental Health will only increase that problem. Across the country, Patricia Weathers, the Carroll Family, the Johnston Family, and the Salazar Family were all

charged or threatened with child abuse charges for refusing or taking their children off of psychiatric medications.

(14) The United States Supreme Court in Pierce versus Society of Sisters (268 U.S. 510 (1925) held that parents have a right to direct the education and upbringing of their children.

(15) Universal or mandatory mental health screening violates the right of parents to direct and control the upbringing of their children.

(16) Federal funds should never be used to support programs that could lead to the increased over-medication of children, the stigmatization of children and adults as mentally disturbed based on their political or other beliefs, or the violation of the liberty and privacy of Amercians by subjecting them to invasive "mental health screening" (the results of which are placed in medical records which are available to government officials and special interests without the patient's consent.)

SEC. 3. PROHIBITION AGAINST FEDERAL FUNDING OF UNIVERSAL OR MANDATORY MENTAL HEALTH SCREENING

(a) Universal or Mandatory Mental Health Screening Program – No Federal funds may be used to establish or implement any universal or mandatory mental health, psychiatric, or socioemotional screening program.

(b) Refusal to Consent as Basis of a Charge of Child Abuse of Education Neglect – No Federal education funds may be paid to any local educational agency or other instrument of government that uses the refusal of a parent or legal guardian to provide express, written, voluntary, informed consent to mental health screeing for his or her child as the basis of a charge of child abuse, child neglect, medical neglect, or education neglect until the agency or instrument demonstrates that it is no longer using such refusal as a basis of such a charge.

(c) Definition – For purposes of this Act, the term "universal or mandatory mental health, psychiatric, or socioemotional screening program"

(1) means any mental health screening program in which a set of individuals (other than members of the Armed Forces or individuals serving a sentence resulting from conviction for a criminal offense) is automatically screened without regard to whether there is prior indication of a need for mental health treatment; and

(2) includes –

(A) any program of State incentive grants for transformation to implement recommendations in the July 2003 report of the President's New Freedom Commission on Mental Health, the State Early Childhood Comprehensive System, grants for TeenScreen, and the Foundations for Learning Grants; and –

(B) any student mental health screening program that allows mental health screening of individuals under 18 years of age without the express, written, voluntary, informed consent of the parent or legal guardian of the individual involved.

NOTES

CHAPTER 1: SELLING LEGAL DRUGS

Marketing and administration costs of ten largest firms reported in "Pill Pushers" by Robert Langreth and Matthew Herper, *Forbes*, May 8,2006, p.97.

Cases of fraud reported in; "Pushing the Pills a Bit Too Hard" by Arlene Weintraub, *BusinessWeek* Feb.26, 2007, p.48.

Drug price increase reported in *AARP* news release, June 21, 2006.

Medicare drug price increase reported in, "Cost Grow for Common Medicare Drugs", by Jonathan Weisman, *Washington Post*, May 13, 2007.

Bankruptcy rate reported in "Healing Our System" by Patricia Barry and Barbara Basler, *AARP Bulletin*, March 2007, p.12.

Number of people on Medicare Part D is from the Department of Health and Human Services (HHS); www.hhs.gov

Number of people qualified to receive government assistance for pharmaceuticals is from ibid; and *USA Today,* August 2005, from http://USATODAY.com.

Money spent in the U.S. for prescription drugs in 2006 is from the FDA website, http://www.fda.gov and the website http://www.wikipedia.org.

Public money spent on prescription drugs is from Report by California Healthline, http://www. californiahealthline.org.

U.S. per capita number of prescriptions in 2005 is from the Kaiser Family Foundation website http://www.kff.org.

Harvard study about lifestyle affecting health is from http://www.newstarget.com/z020397.html.

People without health insurance, "Divided We Fall", *AARP Bulletin*, Feb.2007, p.3.

Illinois farmer is from the Illinois governor's website http://illinois.gov.

Ray Moynihan and Alan Cassels, *Selling Sickness, How the World's Biggest Pharmaceutical Companies Are Turning Us All Into Patients* (New York:Nations Books, 2005).

Elderly not taking prescription drugs, www.aarp.org and www.kff.org.

Henry Gadsden quote, Ray Moynihan and Alan Cassels, *Selling Sickness, How the World's Biggest Pharmaceutical Companies Are Turning Us All Into Patients* (New York:Nations Books, 2005), p.ix;. and article by W.Robertson, *Fortune*, March 1976.

CHAPTER 2: THE HISTORY OF MARKETING PHARMACEUTICALS

Number of drug representatives reported in "Pill Pushers" by Robert Langreth and Matthew Herper, *Forbes*, May 8,2006, p.97; and www.verispan.com.

Timeline of changes in marketing is from Kay Carlson's interviews with pharmaceutical representatives that have worked in the industry since the 1960's.

U.S. prescriptions in 2005, "Prescription Drug Trends" June 2006, Kaiser Family Foundation, http://www.kff.org.

Direct to consumer advertising costs, ibid.

CHAPTER 3: EXPENSIVE DOOR-TO-DOOR SELLING TO DOCTORS

One drug rep. for every 9 physicians was reported "Pill Pushers" by Robert Langreth and Matthew Herper, *Forbes*, May 8,2006, p.97.

Triple number of drug representatives is from www.verispan.com.

Billions spent in 2005 for free drug samples is from "Thanks, But No Thanks," by Anne Underwood, *NewsWeek*, October 29, 2007, p.49.

"Health Facilities Flush Estimated 250M Pounds of Drugs a Year," by Jeff Don, Martha Mendoza and Justin Pritchard, *USA Today*, September 14, 2008.

Eric G. Campbell, Ph.D. et al, "A National Survey of Physician-Industry Relationships", *New England Journal of Medicine*, 356:1742-1750, April 26, 2007.

Doctors beginning to say no to drug representatives is from "Thanks, But No Thanks," by Anne Underwood, *NewsWeek*, October 29, 2007, p.49.

CHAPTER 4: TRICKS TO PAY DOCTORS TO WRITE PRESCRIPTIONS

Gene Carbona reported in "After Sanctions, Doctors Get Drug Company Pay" by Gardiner Harris and Janet Roberts, *New York Times*, June 3,2007.

Oncologists paid for drug use from "Doctors Reaping Millions for Use of Anemia Drugs" by Alex Berenson and Andrew Pollack, *New York Times*, May 9, 2007.

FDA reports and warnings for EPO drugs is from http://www.fda.gov.

CHAPTER 5: GUIDELINES ARE MADE FOR DRUG SALES, NOT HEALTH

Dr. George V. Mann, "Doing the Wrong Things," *Nutrition Today*, vol. 58: 423-428, June 1985.

"Coronary Heart Disease: The Dietary Sense and Nonsense," edited by George V. Mann, M.D., (New York; Veritas Society, 1993).

Guidelines from Ray Moynihan and Alan Cassels, *Selling Sichness, How the World's Biggest Pharmaceutical Companies Are Turning Us All Into Patients* (New York:Nations Books, 2005), pp.7-8.

Dr. Uffe Ravnskov quote is from "Cholesterol Skeptics and the Bad News About Statin Drugs," by Maryann Napoli, http://www.medicalconsumers.org/pages/cholesterol_skeptics.html.

Lipitor sales is from http://money.cnn.com /2006/08/02/news/companies/lipitor.

ALLHAT study, http://nhlbi.nih.gov/health/allhat /furberg.htm and http://jama.ama-assn.org/.

Dr. Abramson quote from Ray Moynihan and Alan Cassels, *Selling Sichness, How the World's Biggest Pharmaceutical Companies Are Turning Us All Into Patients* (New York:Nations Books, 2005), p.8.

Statins not beneficial to women is from the University of British Columbia Therapeutics Initiative website http://www.ti.ubc.ca/en/node/52

Scandinavian Simvastatin Survival Study Group, *Lancet*, vol. 344: 1383-1389, 1994.

Statin sales is from http://seniorjournal.com /NEWS/HEALTH/2008/8-01-29-senCitStunned.html

Dr. Paul Rosch quote is from "Cholesterol Skeptics and the Bad News About Statin Drugs," by Maryann Napoli, http://www.medicalconsumers.org/pages/cholesterol_skeptics.html.

Dr. Duane Graveline's quote is from www.spacedoc.net.

French study of women is from "Cholesterol Skeptics and the Bad News About Statin Drugs," by Maryann Napoli, http://www.medicalconsumers.org/pages/cholesterol_skeptics.html.

European study of men and women is from "Yet Another Study Shows Low Cholesterol Increases Risk of Early Death!" by Chris Gupter, www.newmediaexplorer.org /chris/2006/06/08.html.

Hawaiin study is from "Cholesterol and all-cause mortality in elderly people from the Honolulu Heart Program: a cohort study," by Irwin J. Schatz, M.D. et al, *The Lancet*, vol. 358, no. 9279, pp. 351-355.

Dr. Mary Enig's quote is from "Dangers of Statin Drugs: What You Haven't Been Told About Cholesterol-Lowering Medicines," by Mary Enig, Ph.D. and Sally Fallon, www.westonaprice.org/moderndiseases/statin.html.

The Jupiter study is from "Rosuvastatin to Prevent Vascular Events in Men and Women with Elevated C-Reactive Protein," by Paul M. Ridker, M.D. et al, *New England Journal of Medicine,* 359: 2195, November 20, 2008.

Dr. Nieca Godberg quote is from "A Call for Caution in the Rush to Statins," by Tara Parker-Pope, *The New York Times,* November 18, 2008.

CHAPTER 6: FRAUDULENT MARKETING CAUSES DEATHS

Harry Loynd quote is from Donald L. Barlett and James B. Steele, *"Critical Condition – How Healthcare in America Became Big Business and Bad Medicine,"* (New York, Doubleday, 2004) p. 233.

Pfizer 2009 settlement is from "In Settlement, A Warning to Drugmakers," by Carrie Johnson, *Washington Post,* September 3,2009.

Cases of alleged fraud from "Pushing the Pills a Bit Too Hard" by Arlene Weintraub, *BusinessWeek,* February 26, 2007.

OxyContin case from "Narcotic Maker Guilty of Deceit Over Marketing" by Barry Meier, *New York Times,* May 11, 2007; and "OxyContin Maker, Execs Guity of Deceit" by Sue Lindsey, *Associated Press,* http://news.yahoo.com/s/ap/20070511/ap_on_bi_ge/oxycontin_plea.

Plavix from http://advisor.investopedia.com/news /06/Bristol-Myers_Hurt-By-Plavix/; and "Bristol Myers Agrees to Plead Guilty in Plavix Case" by Stephanie Saul, *New York Times,* May 11, 2007.

Cost of Plavix is from Kay Carlson's interview with pharmacist Tim Clark.

Bristol-Myers Squibb pays $515 million is from http://www.whistleblowerlawyerblog.com/2007/10/bristolmeyers_squibb _pays_515.

NIH whistleblower Dr. Susan Molchan is from www.ahrp.org/cms/content/view/282/27

Aricept sales is from http://money.cnn.com/2007 /08/06/news/companies/myrad/index.html

Pfizer admits guilt in promotion of Neurontin is from http://www.ahrp.org/infomail/04/05/16.php

Pfizer fraud is from "Huge Penalty in Drug Fraud Pfizer Settles Felony Case in Neurontin Off-Label Promotion," by Bernadette Tansey, *San Francisco Chronicle,* May 14, 2004.

TAP Pharmaceutical settled criminal charges is from http://www.psa-rising.com/wirebird/tap102001.php and http://www.insure.com/articles/lawsuitlibrary/tap_pharmaceuticals.html.

Schering-Plough settlement is from
http://www.boston.com/articles/2006/08/30/drug_firm_hit_with_3d_big_
penalty_in_five_years.

Sanofi-Aventis settlement is from
http://www.usdoj.gov/opa/pr/2007/September/07_civ_694.html.

Zyprexa lawsuits from http://www.law.com/jsp/law; and
www.ahrp.org; and http://www.lawyersan
dsettlements.com/articles/zyprexa_lawsuit.html

Zyprexa 2006 sales is from "Lilly Receives Zyprexa Greetings From
Capitol Hill," by Evelyn Pringle, April 27, 2007,
http://www.lawyersandsettlements.com/zyprexa_lawsuits.html.

Thomas J. Moore, et al, "Serious Adverse Drug Events Reported to the
Food and Drug Administration, 1998 – 2005", *Archives of Internal Medicine,*
167 (16):1752-1759, September 10, 2007.

Congress investigating Eli Lilly is from "Disparity Emerges in Lilly
Data on Schizophrenia Drug," by Alex Berenson, *New York Times,* December
21, 2006; and "Lilly Recieves Zyprexa Greetings from Capitol Hill," by
Evelyn Pringle, April 27, 2007;
http://www.lawyersandsettlements.com/articles/00805/zyprexa-capitol-
hill.html

Death sentenance in China from Asia-Pacific News, May 29, 2007.

Kurt Eichenwald, "*Conspiracy of Fools*", (New York, Broadway Books,
2005), p.10.

CHAPTER 7: REPORTING SERIOUS SIDE EFFECTS

Peter Wilmshurst, "Heart Protection Study," *Lancet,* vol. 361: 528-529,
February 8, 2003.

Thomas J. Moore, et al, "Serious Adverse Drug Events Reported to the
Food and Drug Administration, 1998 – 2005", *Archives of Internal Medicine,*
167 (16):1752-1759, September 10, 2007.

Drugs withdrawn 1997 – 2000 from http://www.gao.gov, "GAO-01-
286-R Drugs Withdrawn From Market".

Survey from Integrity in Science at http://www.cspinet.org/
integrity/watch/200607243.html.

Scientists sued is from "Big Pharma, Bad Science" by Nathan Newman,
July 25, 2002, *The Nation,* at http://www.thenation.com/doc/
20020805/newman20020725.

Betty Dong, ibid.

Drugs withdrawn 2000-March 30,2007 is from
http://www.answers.com/topic/list-of-withdrawn-drugs;
and http://fda.gov/FDAC/features/2002/102_drug.html.

Higher risk to women from http://www.gao.gov, "GAO-01-286-R
Drugs Withdrawn From Market".

Number of drugs on U.S. market is from http://www.fda.gov

Women heart attack deaths is from University of Pennsylvania Health
System website, http://pennhealth.com; and *Circulation – Journal of American
Heart Association* website, http://circ.ahajournals.org/cgi.

Asians on statins is from my training as a pharmaceutical
representative.

Texas Attorney General vs. Janssen is from "Pushing the Pills a Bit Too
Hard" by Arlene Weintraub, *BusinessWeek,* February 26, 2007, p.48.

CHAPTER 8: POLITICS OF DANGEROUS DRUGS

Number of deaths from Vioxx is from
http://mynippon.com/viox/2005/01/vioxx-death-estimates-revised-
upward.html.

C. Bombardier, M.D., et al, "Comparison of Upper Gastrointestinal
Toxicity of Rofecoxib and Naproxen in Patients with Rueumatoid Arthritis,"
New England Journal of Medicine, vol. 343:1520-1528, November 23, 2000.

Cleveland Clinic study, D. Mukherjee, S. Nissen, and E. Topol, "Risk of
Cardiovascular Events Associated with Selective COX-2 Inhibitors," *Journal
of American Medical Association,* vol.286, August 2001, pp. 954-959.

5 million prescriptions is from "Vioxx: Downfall of a Superdrug," by
Katharine Greider, *AARP Bulletin,* November 2004; on
www.aarp.org/bulletin/prescription/a2004- 10-04-vioxx.html.

Merck internal memo is from http://www.mynippon.com/vioxx
/2005/05/merck-sold-vioxx-ruthlessly.html. Made available by House
Reforms Committee.

FDA warning letter to Merck is from "Vioxx: Downfall of a
Superdrug," by Katharine Greider, *AARP Bulletin,* November 2004.

Administrative law is from Kay Carlson's interview with Leon
Santman, U.S. Department of Transportation, Assistant General Counsel,
retired.

Political contributions is from
http://www.mynippon.com/vioxx/2005/05.

Vioxx voluntarily withdrawn is from http://www.fda.gov.

Dr. Graham's study is from Vioxx: Downfall of a Superdrug," by Katharine Greider, *AARP Bulletin,* November 2004.

FDA warning Merck is from Integrity In Science, http://www.cspinet.org/integrity/watch/200607243.html

February 2005 FDA advisory committee meeting is from Kay Carlson's interview with Merrill Goozner, Director, Center for Science in the Public Interest, May 9, 2007.

Evaluation of committee members is from http://www.cspinet.org/new/200502251.html.

COX-2 Inhibitor studies is from http://www.cspinet.org/integrity/press/200502161.html.

Merck settlement of lawsuits is from http://enews.earthlink.net/channel/news.

FDA letter to Pfizer is from http://www.mynippon.com/vioxx/01/fda-warns-pfizer-on-misleading.html.

Bextra withdrawal is from http://www.fda.gov and http://www.mayoclinic.com/Health/bextra/AR00042.

CHAPTER 9: MEDICATING THE HEALTHY

Thomas L. Harrison quote is from Donald L. Barlett and James B. Steele, *"Critical Condition – How Healthcare in America Became Big Business and Bad Medicine,"* (New York, Doubleday, 2004) p. 229.

"The Lifestyle Drugs Outlook to 2008: Unlocking New Value in Well-Being," *Reuters Business Insight,* October 2003, http://www.globalbusinessinsights.com/rbi/report.asp?id=rbhc0109.

Deaths due to prescription drugs is from Barbara Starfield et all, "Is U.S. Healthcare Really the Best in the World?", Journal of American Medical Association, 284:483-485, July 2000.

FDA warning letters is from http://www.uspirg.org/home/reports/report-archives/healthcare/health

Harry Loynd quote is from Donald L. Barlett and James B. Steele, *"Critical Condition – How Healthcare in America Became Big Business and Bad Medicine,"* (New York, Doubleday, 2004) p. 233.

Marketing false claims is from http://www.uspirg.org/home/reports/report-archives/healthcare/health.

St.John's Wart study is from http://nccam.nih.gov/news/2002/stjohswort/Pressrelease.html.

Fosamax study is from Ray Moynihan and Alan Cassels, *Selling Sichness, How the World's Biggest Pharmaceutical Companies Are Turning Us All Into Patients* (New York:Nations Books, 2005), p. 151.

Numbers needed to treat is from my training as a pharmaceutical representative.

Healthy life expectancy rate is from http://www.who.int/choice /publications/d_2000_gpe38.pdf.

Population and drug use is from Ray Moynihan and Alan Cassels, *Selling Sichness, How the World's Biggest Pharmaceutical Companies Are Turning Us All Into Patients* (New York:Nations Books, 2005), p. xi.

CHAPTER 10: CANCER IS BIG BUSINESS

Cancer's worth to medical system is from http://www.mnwelldir.org.

Quotes are from http://curezone.com/diseases /cancer/chemo_ therapy_facts.asp.

Cost of cancer treatment in U.S. is from "Is High and Growing Spending on Cancer Treatment and Prevention Harmful to the United States Economy?" by Mark V. Pauly, *Journal of Clinical Oncology*, vol. 25:2, January 10, 2007, pp. 171-174.

Cost of Avastin and Terceva, ibid.

Cost of Tyderb is from "U.S. Approves One-A-Day Pill for Advanced Breast Cancer," by Associated Press, *The Boston Globe*, March 14, 2007.

Quote from Dr. Ralph Moss is from an interview with Laurie Lee on British television, 1994.

About Dr. Moss is from http://www.cancerdecisions .com /about.html

Amygdalin facts are from http://en.wikipedia .org/wiki/Amygdalin.

Cat's Claw research findings is from http://Amazondreams.amazonherb.net.

Nordic Cochrane Centre findings of mammography screening is from "Dissenting Think in Pink Ribbon Week," by Michael Woodhead, *6 Minutes of Interesting Stuff for Doctors Today*, October 23, 2007, http://www.6minutes.com.au.

Thomas Edison quote is from http://en.wikipedia.org /wiki/Clarence_Madison_Dally.

Mary Helen Barcellos_Hoff quote is from http://www.preventbc.com/news.html.

German study of low radiation DNA damage is from ibid.

NIH going against its own expert panel advising against mammography screening is from http://www. Sausa.org/lofiversion/ index.php.t2505.html.

Metastasis vs. tumor size is from "Why We're Losing the War on Cancer," by Clifton Leaf, *CNN.com*, January 12, 2007.

Dr. James Oschman's quote is from his keynote address to the University of North Carolina Medical School's CAM conference, spring 2005.

Study from British Medical Journal is from "Why We're Losing the War on Cancer," by Clifton Leaf, *CNN.com*, January 12, 2007.

Taxol sales is from "Once Should Be Enough," by Katharine Greider, *AARP Bulletin,* May 2006.

Erbitux, ibid.

TAP Pharmaceuticals drug fraud charges is from "New Salvos in the Prescription Drug Wars," by Patricia Barry, *AARP Bulletin,* January 2005.

Roland T. Skeel, M.D. and Neil A. Lachant, M.D., *Handbook of Cancer Chemotherapy:* Fourth Edition, (New York, Little, Brown and Company, 1995), p.667.

Cost of chemo adverse effects for breast cancer is from M.J. Hassett, et al, "Frequency and Cost of Chemotherapy-related Serious Adverse Effects in a Population Sample of Women with Breast Cancer," *Journal of the National Cancer Institute,* 98 (16):1108-17, August 16, 2006.

Cost of chemo adverse effects for ovarian cancer is from E.A. Calhoun, et al. "Evaluating the Total Costs of Chemotherapy-induced Toxicity: Results from a Pilot Study with Ovarian Cancer Patients," *Oncologist,* 6(5):441-5.

Chemobrain is from "Chemotherapy Found to Cause Permanent Brain Damage, Loss of Memory," by Ben Kage, *News Target.com,* October 6, 2006.

Hearing Loss is from "Hearing Loss from Chemotherapy Underestimated," *News Target.com,* September 13, 2006.

Hodgkin's survivors is from http://www.mayoclinic.com.

Oncologists would not take chemo but use on 75% of patients is from Dr. James Oschman's keynote address to the University of North Carolina Medical School's CAM conference, spring 2005.

Danny Hauser story is from www.dannyhauser.com

Billy Best story is from www.abcnews.go.com/US/story?

Abraham Cherrix story is from "Judge Orders Abraham Cherrix Into Radiation Treatment for Cancer," *News Target.com,* August 16, 2006.

Katie Wernecke story is from, "Does the State Own Your Body?", by Jessica Fraser, *News Target.com,* August 3, 2006.

Increased risks of chemo and radiation for Hodgkin's disease is from http://mayoclinic.com.

Dr. Lorraine Day's story is from http://www.drday.com.

Gardasil vaccine is from "Virginia Law Guaranteeing Parents' Medical Rights Routed by Mandatory HPV Vaccination," by Ben Kage, *News Target.com,* February 20, 2007.

Cervical cancer information is from Kay Carlson's interview with Myra Hall, M.D.

Merck lobbies states for mandatory vaccinations is from "Yikes! An STD Vaccine for Sixth-graders" by Carolyn Sayre, *Time,* February 8, 2007; www.time.com.

Cost of vaccine is from ibid.

Deaths and side effects of Gardasil is from "Postlicensure Safety Surveillance for Quadrivalent Human Papillomavirus Recombinant Vaccine," *Journal of American Medical Association,* Aug. 19, 2009, 302(7);750-757.

Dr. Scott Ratner's quote is from http://www.cbsnews.com/stories/2009.

Infant mortality is from http://en.wikipedia.org/wiki/list_of_countries_by_infant_mortality_rate.

CHAPTER 11: PEOPLE WITHOUT CONSCIENCE

1997 Congressional resolution is from http://www.answers.com/topic/2000-simpsonwood-cdc-conference.

Sunshine Laws, ibid.

Meeting participants is from "Deadly Immunity," by Robert F. Kennedy, Jr., at http://www.rollingstone.com/politics/story/7395411/deadly_immunity.

Dr. Tom Verstraeten quote is from the Simpsonwood meeting minutes posted on http://www.nomercury.org/science/documents/simpson wood%20overview.pdf

Dr. Bill Weil quote, ibid.

Dr. Robert Brent, quote ibid.

Dr. John Clements quote, ibid.

Dr. Verstaeten warning to CDC and FDA in 1999 is from http://www.answers.com/topic/2000-simpsonwood-cdc-conference.

Dr. Roger Bernier quote is from the Simpsonwood meeting minutes posted on http://www.nomercury.org/science /documents/simpsonwood % 20overview.pdf.

Database given to private company is from "Deadly Immunity," by Robert F. Kennedy, Jr..

Thimerosal laced vaccines shipped to third world countries, ibid.

ACIP changes to vaccinations is from ibid, and http://www.answers.com, and http://www.autismcoach.com.

Pregnant women not safe from Thimerosal is from "Autism – Cut the Crap," by Evelyn Pringle, *The Sierra Times*, updated June 14, 2007, http://www.sierratimes.com /05/07/29/24_164_252_187_13884.html.

Leonard Trasande, MD study of mercury levels in newborns is from "Thimerosal Definate Cause of Autism," by Eveyln Pringle, January 29, 2007, www.scoop.co.nz/stories/HL0503/S00089.htm

David Ayoub, MD quote is from ibid.

Drop in autism in California is from http://www.autismcoach.com.

Contribution to Senator Bill Frist is from "Deadly Immunity," by Robert F. Kennedy, Jr. and http://www.answers.com.

States banning mercury in vaccines is from http://www.a-champ.org/state.html.

Dr. Maurice Hilleman's advice to Merck is from "Deadly Immunity," by Robert F. Kennedy, Jr..

Research results from Mark R. Geier, M.D. and David A. Geier, B.A. "Early Downward Trends in Neurodevelopmental Disorders Following Removal of Thimerosal-Containing Vaccines," *Journal of American Physicians and Surgeons,* vol. 11:1, Spring 2006.

Vaccines that still contain Thimerasol is from http://www.starbulletin.com/2006/05/19/news.

CHAPTER 12: THE PUPPET AND THE VENTRILOQUIST

Sally A. Shumaker, et al, "Estrogen Plus Progestin and the Incidence of Dementia and Mild Cognitive Impairment in Postmenopausal Women: Women's Health Initiative Memory Study," *JAMA*, May 2003; 289:2651-2662.

Menopause congress in Sydney is from Ray Moynihan and Alan Cassels, *Selling Sickness, How the World's Biggest Pharmaceutical Companies Are Turning Us All Into Patients* (New York:Nations Books, 2005), p. 50.

The ability to create new disease markets, ibid.; and originates from J.Coe, "Healthcare: The lifestyle drugs outlook to 2008, unlocking new value in well-being," *Reuters Business Insight,* Datamonitor, PLC, 2003.

Sharon Batt quote is from "Unscientific Depression Screenings and Front Groups Boost SSRI Sales," by Evelyn Pringle, http://www.lawyersandsettlements.com/articles/00436/ssri.html.

Racketeering lawsuit against Eli Lilly is from "States Try to Limit Drugs in Medicaid, But Makers Resist," by Gardiner Harris, *New York Times,* December 18, 2003; and "The Secrets in Eli Lilly's Cabinet," by Evelyn Pringle, *Sierra Times,* http://www.sierratimes.com/07/01/20/Pringle.html;

Stephen Hulley, et al, "Randomized Trial of Estrogen Plus Progestin for Secondary Prevention of Coronary Heart Disease in Postmenopausal Women," *JAMA,* August 1998; 280:605-613.

Society for Women's Health Research is from the Center for Science in the Public Interest website http://www.cspinet.org/integrity/nonprofits/society_for_women_146_s_health_research.html.

Heidi D. Nelson, et al, "Postmenopausal Hormone Replacement Therapy: Scientific Review," *JAMA,* August 2002; 288:872-881.

Thomas J. Moore, et al, "Serious Adverse Drug Events Reported to the Food and Drug Administration, 1998 – 2005", *Archives of Internal Medicine,* 167 (16):1752-1759, September 10, 2007.

Wyeth sales of HRT drop is from "Fight brews over natural vs. synthetic hormone treatments," by Kate Nolan, The Arizona Republic, June 26, 2006; http://azcentral.com. arizonarepublic/news/articles/0626hormones0626.html.

Wyeth petition to FDA, ibid.

Warnings are from http://www.fda.gov and http://www.newsinferno.com.

Society for Women's Health Research is from the Center for Science in the Public Interest website http://www.cspinet.org/integrity/nonprofits/society_for_women_146_s_health_research.html.

Zelnorm withdrawn from market is from http://www.fda.gov.

Depression is real campaign is from "Unscientific Depression Screenings and Front Groups Boost SSRI Sales," by Evelyn Pringle, http://www.lawyersandsettlements.com/articles/00436/ssri.html.

Dr. Fred A. Baughman, Jr. quote is from his website http://www.adhdfraud.com

Jacob S. Hacker, *The Great Risk Shift, The Assault on American Jobs, Families, Healthcare, and Retirement And How You Can Fight Back,* New York: Oxford University Press, 2006, p.29 – 32.

U.S. poverty level is from http://aspe.hhs.gov/poverty/07poverty.shtml.

CHAPTER 13: NO VALUE ADDED WHILE DRUG PRICES SKYROCKET

CEO compensations are from http://www.aflcio.org, http://www.sec.gov and http://www.reuters.com.

Drug markups analysis done by Sharon Davis and Mary Palmer, U.S. Department of Commerce, http://www.rense.com/general54 preco.html.

Research spending in 2004, http://en.wikipedia.org/wiki/harmaceutical#drug_info.

$30 billion spent on promotions is from "A Decade of Direct-to-Consumer Advertising of Prescription Drugs," by Julie M. Donohue, Ph.D., Marisa Cevasco, B.A., Meredith B. Rosenthal, Ph.D., *New England Journal of Medicine,* Vol. 357: 673-681, August 16, 2007.

Dr. Meredith Rosenthal's quote if from University of Pittsburgh Schools of the Health Sciences Media Relations, http://www.upmc.com/communications/mediarelations/newsreleasearchives/2007/August

Illinois website for Canadian and English pharmacies is from "States Defy FDA on Drug Importation," by Patricia Barry, *AARP Bulletin,* October 2004.

South Africa's HIV statistics and attempts to treat the epidemic is from http://en.wikipedia.org/wiki/ Pharmaceutical_Drug_information.

CHAPTER 14: GOLD MINES

Money grossed by 5 leading psychiatric drugs is from the documentary film "Making a Killing,", by Citizens Commission on Human Rights.

Dr. Ron Leifer quote is from http://www.cchr.org

Amount spent on psychiatric drugs is from the documentary film "Where the Truth Lies", by Citizens Commission on Human Rights.

Senator Grassley's investigations is from "Psychiatric Group Faces Scrutiny Over Drug Industry Ties," by Benedict Carey and Gardiner Harris, *The New York Times,* July 12, 2008.

Dr. Sharfstein's quote is from "MH System Reform Must Start with Funding," by Kate Mulligan, *Psychiatric News,* Vol. 38, Number 1, January 3, 2003, p. 9.

Texas alliance is from Allen Jones' whistle-blower article on http://psychrights.org.

Grant from J&J is from ibid.

FDA approval letter to Janssen for Risperdal, ibid.

Lawsuits against Eli Lilly for Zyprexa is from http://www.law.com/ jsp/law/ LawArticleFriendly.jsp?id=1175517537442.

Dr. Peter Weidman's quote is from "Guidelines for Schizophrenia: Consensus or Confusion?" by Dr. Weidman, *Journal of Practical Psychiatry and Behavioral Health,* January 1999.

Dr. Daniel J. Carlat quote is from "No to Drug Money, " by Carey Godberg, *The Boston Globe,* May 7, 2007.

Dr. Jerome P. Kassirer's quote is from ibid.

Dr. Irwin Savodnik's quote is from http://www.ahrp.org/cms/ content/view/143/27. Originally from *The Chicago Tribune.*

Lisa Cosgrove, et al, "Financial Ties Between DSM-IV Panel Member and the Pharmaceutical Industry," *Psychotherapy and Psychosomatics,* 2006; 75:154-160.

Minnesota psychiatrists is from "After Sanctions, Doctors Get Drug Company Pay," by Gardiner Harris and Janet Roberts, *The New York Times,* June 3, 2007.

FDA data for deaths in clinical trials is from Allen Jones' whistleblower article, http://psychrights.org.

Dr. John Geddes study, ibid.

Cost of NIMH study is from "Drug Companies Still Peddling Risperdal and Zyprexa for Off-Laabel Use," by Evelyn Pringle, June 17, 2006, http://www.Lawyerandsettlements.com/articles/zyprexa_Off_Label.html.

FDA approval of Risperdal for autistic children is from http://www.fda.gov/dcer/meeting/medication_Guides/05Hasnner Sharav.pdf

FDA approval of Risperdal for bipolar children is from http://www.fda.gov/bbs/topics/news/2007/new01686.html.

J.A. Lieberman, et al, "Effectiveness of Antipsychotic Drugs in Patients with Chronic Schizophrenia," *New England Journal of Medicine, O* 22, 2005, 353:10291233.

2008 antipsychotic sales is from "The Mothers Act Disease Mongering Campaign – Part 1," by Evelyn Pringle, July 16, 2009, www.naturalnews.com/02634_drugs_suicide_ adhd.html.

Dr. David Healey's quote is from Allen Jones' whistleblower article on http://psychrights.org.

Lilly lawsuit settlements is from "Lilly Settles With 18,000 Over Zyprexa," by Alex Berenson, *The New York Times,* January 5, 2007.

VA spending for atypical antipsychotic drugs is from Robert Rosenheck, MD, et al, "Effectiveness and Cost of Olanzapine and Haloperidol in the Treatment of Schizophrenia: A Randomized Controlled Trial," *Journal of American Medical Association*, Nov. 2003, 290: 2693-2702.

Lon S. Schneider, Karen S. Dagerman, Philip Insel, "Risk of Death with Atypical Antipsychotic Drug Treatment for Dementia: Meta-analysis of Randomized Placebo-Controlled Trials," *JAMA,* October 19, 2005, 295:1934-1943.

John Nash story is from Sylvia Nasar, *"A Beautiful Mind,"* (New York:Simon and Schuster,1998)

Soteria Project is from Robert Whitaker, *"Mad in America,"* (Cambridge, MA: Perseus Publishing, 2002) pp. 220-224.

Courtenay Harding, Ph.D. research is from "Forget What the Sceptics Say, Most People With Schizophrenia Get Better," by Judith Carrington, *New York City Voices*, October-December 2003, http://www.newyorkcityvoices.org/2003octdec/20031225.html.

Assen Jablensky, "Schizophrenia: Manifestations, Incidence and Course in Different Cultures, a World Health Organization Ten-Country Study, *Psychological Medicine,* (1992) supplement 20: 1-95.

Drugs banned in Britain is from http://www.ablechild.org/ewsarchive/will_british_ban_spur_fda_to_act%202-2-04.html.

FDA alert for serotonin syndrome is from http://www.fda.gov.

CHAPTER 15: ANTIDEPRESSANTS CREATE KILLERS

Dr. Arif Khan study results is from http://www.ahrp.org/infomail/04/06/10.php.

Black Box Warning if from http://www.fda.gov

Story of Candace's death is from Kay Carlson's interview with Candace's mother.

Story of Aaron Todovich's death is from Kay Carlson's interview with Aaron's mother.

Story of Shawna Scantlin's death is from Kay Carlson's interview with Shawna's husband.

Dr. Charles Nemeroff is from http://www.cchr.org.

German licensing authority letter about Prozac suicides in clinical trials is from "They Said It Was Safe" by Sarah Boseley, *The Guardian*, Oct. 30, 1999.

Internal Lilly memos are from http://www.ablechild.org/newsarchive/will_british_ban_spur_fda_to_act%202-2-04.html.

Lilly's chief scientist quote is from "They Said It Was Safe", by Sarah Boseley, *The Guardian*, Oct. 30, 1999.

Dr. Paul Leber's quote is from "Warnings About a Miracle Drug," by Anastasia Toufexis, *Time*, July 30, 1990.

Dr. David Graham's quote is from http://www.fda.gov/cder/meeting/medication_guidelines/05HasnnerSharav.pdf.

Dr. Thomas Laughren's quote is from ibid.

Seniors commit suicide taking SSRI drugs is from "Big Pharma Bankrupting US Healthcare System," by Evelyn Pringle, August 29, 2006, http://www.lawyersandsettlements.com/articles/00295/big_Pharma_bankrupt.html.

Dr. Christina Spjut quote is from http://www.dangerousmedicine.com.

Joanna Moncriff and Irving Kirsch, "Re-examining SSRIs for Depression," *British Medical Journal*, 331: 155-157, July 16, 2005.

Celexa clinical trials are from "Weighing Benefits of SSRIs Against Suicide Risk," by Evelyn Pringle, December 8, 2006, http://www.lawyersandsettlements.com/article/00489/ssri-suicide-risks.html.

Prozac trials are from ibid.

Dr. Peter Breggin quoted is from http://www.huffingtonpost.com/dr-peter-breggin/the-real-mental-health-1_b_46327.html.

FDA approval of Sarafem is from http://www.fda.gov/bbs/topics/answers/ans01024.html.

Dr. Paula Caplan reference is from Ray Moynihan and Alan Cassels," *Selling Sichness, How the World's Biggest Pharmaceutical Companies Are Turning Us All Into Patients,*" (New York:Nations Books, 2005) p.100.

N.Y. sues GSK is from "Spitzer Sues a Drug Maker, Saying It Hid Negative Data," by Gardiner Harris, *The New York Times*, June 3, 2004.

Victor Motus' story is from "Pfizer Is Not Liable in Suicide Risk Lawsuit," *The New York Times*, December 25, 2001; and http://www.zoloft-suicide-side-effects.com/pgs/zoloft-lawsuits.html and http://www.teenscreentruth.com;

Internal Pfizer document about Zoloft is from http://www.healyprozac.com/Book/Chapter10.pdf.

Dr. Peter Breggin's quote is from his web site http://breggin.com

Dr. John Zajecka quote is from http://www.cchr.org /index.cfm/6984

Dr. Miki Bloch quote is from ibid.

Dr. Ann Blake Tracey interview is from
http://www.drugawareness.org.

Violent acts associated with antidepressants are from
http://en.wikipedia.org/wiki/antidepressants_ and_shootings, and
http://www.modbee.com.

Dr. Ann Blake Tracey qutoe is from http://www.drugawareness.org.

Dominique Slater's death is from http://www.modbee.com.

Antidepressant prescriptions to children and teens is from
http://www.cchr.org.

Ohio babies on drugs is from "Even Babies Getting Treated as Mentally
Ill," by Encarnacion Pyle, *Columbus Dispatch*, April 25, 2005.

Dr. Ellen Bassuk quote is from ibid.

Christina D. Chambers, et al., "Selective Serotonin-Reuptake Inhibitors
and Risk of Persistent Pulmonary Hypertension of the Newborn," *New
England Journal of Medicine*, Vol. 354: 579-587, No. 6, February 9, 2006

Sabrina F. Lisboa, et al., "Behavioral Evaluation of Male and Female
Mice Pups Exposed to Fluoxetine During Pregnancy and Lactation," *Karger's
Pharmacology*, Vol. 80: 49-56, No. 1, 2007, http://content.karger.com.

Senator Robert Menendez receiving money from pharmaceutical
companies is from "The Mothers Act Disease Mongering Campaign – part
1," by Evelyn Pringle, July 16, 2009,
www.naturalnews.com/026634_drugs_suicide_adhd.html.

Texas Medicaid children is from "Big Pharma Bankrupting US
Healthcare System," by Evelyn Pringle, August 29, 2006,
http://www.lawyersandsettlements.com
/articles/00295/big_Pharma_bankrupt.html.

Ohio children on drugs is from "Even Babies Getting Treated as
Mentally Ill," by Encarnacion Pyle, *Columbus Dispatch*, April 25, 2005.

Tennessee children on drugs is from "The Secrects in Eli Lilly's
Cabinet," by Evelyn Pringle, *Sierra Times*, January 20, 2007.

Massachusetts children on drugs is from "Prevalence of Drugs For DSS
Wards Questioned," by Jessica E. Vascellaro, *Boston Globe*, August 9, 2004.

Florida children is from "Biggest Off-Label Drug Marketing Scheme in
US History Part 1," by Evelyn Pringle, November 30, 2006,
http://www.lawyersandsettlements.com /articles/00471/ssri-off-label-
scheme.html.

Dr. Tony Appel's quote is from NBC News on
http://www.youtube.com/watch?v=1SFPJL66p4c

CHAPTER 16: THE SCAM OF MENTAL HEALTH TESTING IN SCHOOLS

New Freedom Commission, http://www.mentalhealthcommission.gov.

American Association of Physicians and Surgeons quote is from AAPS website, http://www.aapsonline.org/nod.

Dr. Jane Pearson quote is from "Inside TeenScreen," by Sandra Lucas, *Idaho Observer*, January 2006.

Colorado teens TeenScreen found suicidal is from http://www.psychsearch.net/Colorado.

Nine out of ten children put on drugs is from "Inside TeenScreen," by Sandra Lucas, *Idaho Observer*, January 2006, originates from *Journal of the American Academy of Child and Adolescent Psychiatry, 2002*.

Teenage suicide rate is from http://www.cdc.gov.

Study finding suicide rate of 52 percent is from "Doctors Ignore Black Box Warnings on SSRIs, " by Evelyn Pringle, April 2, 2007, http://www.lawyersandsettlements.com /Articles/00707/ssri-doctors.html.

Grant money is from http://www.samhsa.gov/grants /2005/nofa/sm05009_mht_sig.aspx, and http://pn.psychiatryonline.org/ chi/content/full/41/9/1-a.

U.S. suicide rate is from http://www.familyfirstaid.org/suicide.html.

Deaths due to drug adverse events is from "The Leading Cause of Death in the U.S. is the Healthcare System," by Gary Null, Ph.D. et al, http://www.angelfire.com/az /sthurston/Leading_Cause_of_Death_in_the_US.html.

Dr. Marcia Angell's quote is from http://en.wikipedia.org/wiki/TeenScreen.

TeenScreen in 43 states is from http://www.ablechild.org.

TeenScreen trying to get around parental consent is from "Inside TeenScreen," by Sandra Lucas, *Idaho Observer*, January 2006.

Rep. Ron Paul's speech to the House is from http://www.house.gov/paul/legis.shtml.

Senator Reid hosting TeenScreen dinner is from http://www.ablechild.net/forum/viewtopic.php? f=7&t.

Chelsea Rhoades' story is from "Inside TeenScreen," by Sandra Lucas, *Idaho Observer*, January 2006.

Aliah Gleason story is from "Medicating Aliah," by Rob Waters, Mother Jones, May/June 2005, http://www.motherjones.com/news/feature/2005/05/medicating_aliah

Colorado youth homeless shelter is from http://www.psychsearch.net/Colorado.

Dr. John Breeding quote is from "Inside TeenScreen," by Sandra Lucas, *Idaho Observer*, January 2006.

Robert Whitaker, "Anatomy of an Epidemic: Psychiatric Drugs and the Astonishing Rise of Mental Illness in America, *Ethical Human Psychology and Psychiatry*, 7: 23 -33, Spring 2005.

Dr. Elizabeth Roberts quote is from the *Washington Post*, October 8, 2006.

Barry Tuner quote is from "Best Kept Secret - SSRIs Do Not Work," by Evelyn Pringle, March 3, 2007, http://www.lawyersandsettlements.com/articles/00642/ssri-secret.html.

Dr. Diller quote is from "Public Has Right To Know Secrets Revealed in Zyprexa Documents," by Evelyn Pringle, *OpEdNews*, January 12, 2007, http://www.opednewscom/articles/genera_evelyn_p_070112_public_has_r ight_to_.htm.

CHAPTER 17: U.S. - LAND OF DRUGGED CHILDREN

NIH consensus statement is from "Death From Ritalin, The Truth About ADHD," by Lawrence Smith, http://www.ritalindeath.com

Dr. Fred A. Baughman, Jr. quote is from his website http://www.adhdfraud.com.

Daytrana's FDA approval is from the Carlat Psychiatry Report by Daniel J. Carlat, M.D., http://thecarlatreport.com.

Dr. Peter Breggin quote is from "Dr. Breggin Testifies Before U.S. Congress," by Dr. Breggin, September 29, 2000, http://www.breggin.com/classactionmore.html.

Steve Plog's experience with CHADD is from "Making a Killing," a documentary film from Citizen's Commission on Human Rights, http://www.cchr.org.

Inhumane research is from Robert Whitaker, "*Mad in American*",(Cambridge, MA,: Perseus Publishing, 2002) pp. 239-242.

Death rate of legal drugs is from "The Leading Cause of Death," by Gary Null, Ph.D., et al, http://www.angelfire.com/az/sthurston/Leading_Cause_of_Death_in_the _US.html.

DEA listed complications is from "Resolution to Ban Ritalin," by Patti Johnson, presentation to Colorado State Board of Education,http://www.nfgcc.org/banritalin.htm.

Federal Education Department paying schools for ADD/ADHD is from "Resolution to Ban Ritalin," by Patti Johnson, presentation to Colorado State Board of Education, http://www.nfgcc.org/banritalin.htm.

Eight million children taking ADHD drugs is from "Testimonials and Case Histories, What's Wrong with Ritalin?", http://www.hperactivekids.com/quicklinds/whatswrongwithritalin.html.

2008 sales of ADD/ADHD drugs is from "The Mothers Act Disease Mongering Campaign – part 1," by Evelyn Pringle, July 16, 2009, www.naturalnews.com/ 026634_drugs_suicide_adhd.html.

Brookhaven Laboratory research is from "Testimonials and Case Histories, What's Wrong with Ritalin?", http://www.hperactivekids.com/quicklinds/whatswrongwithritalin.html.

Prescription drug abuse is from "Under the Counter: The Diversion and Abuse of Controlled Prescription Drugs in the U.S.", Columbia University National Center on Addiction and Substance Abuse, http;//www.Casacolumbia.org/ Absolutenm/articlefiles/380-final_report_not_embargoed.

Matthew Smith's story is from "Death from Ritalin, The Truth Behind ADHD," by Lawrence Smith, http://www.ritalindeath.com.

Stephanie Hall's story is from "ADHD – Exposing the Fraud of ADD and ADHD," by Dr. Fred A. Baughman, Jr., http://www.adhdfraud.com.

FDA declined Black Box Warning is from http://en.wikipedia.org/wiki/Methylphenidate.

FDA warning is from http://www.fda.gov.

Canada suspends Adderall sales is from "Senator Says FDA Asked Canada Not To Suspend Drug," by Gardiner Harris and Benedict Carey, *The New York Times,* February 11, 2005.

M.D. Anderson Cancer Center study is from http://en.wikipedia.org/wiki/Methylphenidate.

ADHD drug violence is from "FDA Forgot a Few ADHD Drug Related Deaths and Injuries," by Evelyn Pringle, February 15, 2006, http://www.lawyersandsettlements.com/ articles/00103/adhd_fda.html; and "Ritalin Kids: Prescription Drugs and Murder," by Bruce Wiseman, presented to The Pennsylvania House Democratic Policy Committee, http://www.oikos.org/ ritalinkids.html; and "Ritalin and Menticide: America's Opium War Against Its Own Children," by Michele Steinberg, http://www.schillerinstitute.org/ new_viol/michele_article.html.

Diane McGuinness quote is from "The Hazards of Treating 'Attention-Deficit/Hyperactivity Disorder' with Methylphenidate (Ritalin)," by Peter R.

Breggin, M.D. and Ginger Ross Breggin, http://www.breggin.com/methylphen.html.

2005 profit for ADHD drugs is from Citizens Commission on Human Rights, http://www.cchr.org/index.cfm/9027/19552.

Dr. Fred A. Baughman quote is from "ADHD – Exposing the Fraud of ADD and ADHD," by Dr. Fred A. Baughman, Jr., http://www.adhdfraud.com.

Peter R. Breggin, M.D. and Ginger Ross Breggin, "The Hazards of Treating 'Attention Deficit/ Hyperactivity Disorder' with Methylphenidate (Ritalin)," *The Journal of College Student Psychotherapy,* Vol. 10 : 55-72, No. 2, 1995. and http://www.breggin.com/methylphen.html.

Food coloring and sodium benzoate cause hyperactivity is from http://www.medscape.com/viewarticle/562631.

Aspartame is from "Mary Nash Stoddard Research Findings on Dangers of Aspartame at Tesla Conference" http://www.mercola.com/article/aspartame/tesia_conference.htm.

Lead poisoning is from http://www.azsba.org /lead3strange.htm.

Academy of Pediatrics report on lead risk is from Jeanita W. Richardson, "The Cost of Being Poor: Poverty, Lead Poisoning, and Policy Implementation," *The Journal of the American Medical Association,* Vol. 295: 204, April 12, 2006.

Lead poisoning in children is from Bruce P. Lanphear, MD, MPH, "Childhood Lead Poisoning Prevention, Too Little, Too Late," *The Journal of the American Medical Association,* Vol. 293: 2274-2276, May 11, 2005.

Mineral and vitamin supplements improve children's cognitive abilities is from "Nutrition Improves Learning and Memory in School Children," *Medical News Today,* http://www.medicalnewstoday.com.

CHAPTER 18: YOUR HEALTH IS YOUR LIFE

Healthcare spending is from the United Nations Human Development Report,http://www.hdr.undp.org.

HALE is from the World Health Organization, http://www.who.int/choice/publications/d_2000_gpe38.pdf.

Higher U.S. per capita spending is from *AARP Bulletin,* February 2007.

Life expectancy factors is from "Life Expectancy of Many Americans the Same As Citizens of Third-World Countries," *News Target*.com, September 12, 2006, http://www.newstarget.com/z020397.html.

Alcohol self medication for anxiety and depression is from keynote address by Pat Love, Ph.D. at the National Wellness Conference, Stevenspoint, Wisconsin, July 2002.

Ramachandran Vasan, M.D., "Soft Drinks Linked to Metabolic Syndrome," *Circulation,*116:375-384, July 24, 2007.

Dr. Suzanne R. Steinbaum quote is from "Study Links Diet Soft Drinks With Cardiac Risk," by Ed Edelson, http://www.healthfinder.gov/news.

Neal Barnard, M.D. quote is from "How To Solve The Diabetes Epidemic," by Terrence McNally, AlterNet.org, March 14, 2007, http://www.truthout.org/issues_06.

Andrew Weil, M.D. quote is from "Dr. Debunker," by Andrew Weil, M.D., *The AARP Magazine,* May/June 2007, p. 40.

Arctic root is from Andrew Weil, M.D., *Healthy Aging, A Lifelong Guide to Your Physical and Spiritual Well-Being,* (New York: Alfred A. Knopf, 2005), p. 40-41.

Dr. Albert Szent-Gyorgyi quote is from key note address by James L. Oschman, Ph.D. at the Complimentary-Alternative Medicine Conference, UNC Medical School, April, 2005.

Dr. James L. Oschman quote is from ibid.

Molecules do not have to touch is from James L. Oschman, Ph.D., *Energy Medicine, The Scientific Basis,* (London: Elsevier Science Limited, 2002), p. 60.

Water as a cure is from F. Batmanghelidj, M.D., "*You Are Not Sick, You Are Thirsty – Water: For Health, For Healing , For Life,*" (New York: Warner Books, 2003)

Dr. Batmangehelidj quote about stress is from ibid. p. 185.

Dark therapy for bipolar is from "Keeping Bipolar Disorder in the Dark May Just Do the Trick," by Nan Little, August 2, 2007, http://www.InsightJournal.com.

Green light therapy for stress reduction is from http://www.Dr. Emily Kane.com

Joan M. Lappe, et al, "Vitamin D and Calcium Supplementation Reduces Cancer Risk," *American Journal of Clinical Nutrition,* 85: 1586-1591, number 6, June 2007.

Creighton University vitamin D research is from "Creighton Study Shows Vitamin D Reduces Cancer Risk," by Kathryn Clark, http://www2.creighton.edu/publicrelations/newscenter/news/2007/june2007/ June82007/vitamind_cancer_nr060807/index.php.

Canadian Cancer Society recommends vitamin D is from http://www.cancer.ca/ccs/internet/mediareleaselistpf/o3208,3543_434465_1998457305_langld-en,00.html.

Importance of vitamin D is from "Can Vitamin D Help Prevent Pre-Eclampsia?" by Anita Khalek, October 3, 2007, http://www.newstarget.com.

Michael F. Holick, M.D., Ph. D., "Vitamin D Deficiency," *New England Journal of Medicine*, 357 (3): 266-281, July 19, 2007.

Naomi Judd is from "Saved by the Belle," by Joe Nick Patoski, *The AARP Magazine*, May/June 2007, p. 63.

Dr. Stephen Ilardi's program for depression is from "Simply Happy" by Julia M. Klein, *The AARP Magazine*, November/December 2007, p. 50 – 51.

John Lee, M.D. is from Joan Borysenko, Ph.D., *"A Woman's Book of Life, The Biology, Psychology, and Spirituality of the Feminine Life Cycle,"* (New York: Riverhead Books, 1996) p. 175-176; and http://www.womentowomen.com/enopause/estrogendominance.

Natural progesterone is from ibid. p. 170.

Thomas J. Moore, et al, "Serious Adverse Drug Events Reported to the Food and Drug Administration, 1998 – 2005", *Archives of Internal Medicine*, 167 (16):1752-1759, September 10, 2007.

Webster's New World Dictionary of the American Language, Second College Edition, (New York: Simon and Schuster, 1980).

US healthcare system third leading cause of death is from Barbara Starfield,Ph.D. "Is US Health Really the Best in the World?" *Journal of the American Medical Association*, 284: 483-485, July 2000.

Number of people injured by drug errors in hospitals is from "Preventing Medical Errors: Quality Chasm Series," Institute of Medicine, July 20, 2006, http://www.iom.edu./CMS/3809/22526/35939.aspx.

"The Leading Cause of Death in the U.S. is the Healthcare System," by Gary Null, Ph.D. et al,

Index